T0188845

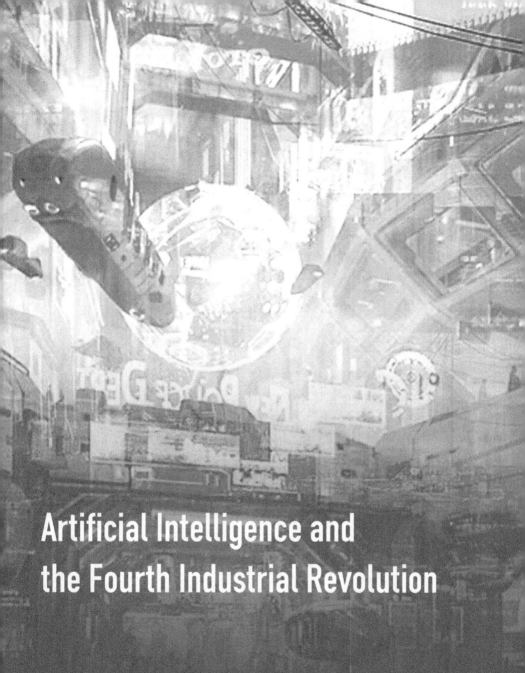

Artificial Intelligence and
the Fourth Industrial Revolution

Artificial Intelligence and the Fourth Industrial Revolution

edited by
Utpal Chakraborty | Amit Banerjee
Jayanta Kumar Saha | Niloy Sarkar
Chinmay Chakraborty

CRC Press
Taylor & Francis Group
Boca Raton London New York

CRC Press is an imprint of the
Taylor & Francis Group, an **informa** business

Artificial Intelligence and the Fourth Industrial Revolution

First published 2022 by Jenny Stanford Publishing Pte. Ltd.

Published 2023 by CRC Press
Taylor & Francis Group
6000 Broken Sound Parkway NW, Suite 300
Boca Raton, FL 33487-2742

ISBN 978-981-4800-79-2 (pbk)
ISBN 978-1-003-15974-2 (ebk)

Visit the Taylor & Francis Web site at
http://www.taylorandfrancis.com

and the CRC Press Web site at
http://www.crcpress.com

British Library Cataloguing-in-Publication Data
A catalogue record for this book is available from the British Library.

Contents

Section II Internet of Medical Things (IoMT)

Preface

This book presents the overall technology spectrum in artificial intelligence (AI) and the fourth industrial revolution that, as we know it, is set to revolutionize the world. The book covers case studies from industry, academics, administration, law, finance and accounting, and educational technology and will be useful for CEOs, entrepreneurs, and university VCs as well as for the vast workforce and students with technical or non-technical backgrounds. The contributing authors are experts in the field and have, from the many interesting topics, focused on gesture recognition prototype for specially abled people, jurisprudential approach to AI and legal reasoning, automated Chatbot for autism spectrum disorder using AI assistance, Big Data Analytics and IoT, role of AI in advancement of drug discovery and development, opportunities and challenges of the fourth industrial revolution, legal ethical and policy implications of AI, Internet of Health Things for smart healthcare and digital well-being, technologies, architecture and opportunities, Internet of Health Things—opportunities and challenges, machine learning and computer vision, computer vision-based system for automation and industrial applications, AI–IoT in home-based healthcare—an effective model for low cost healthcare, and AI in super precision human brain and spine surgery, in this book.

The book covers comprehensive theoretical, methodological, well-established, and validated empirical examples and are of interest for a very vast audience from basic science to engineering and technology experts and learners. It can also be used as a textbook for engineering and biomedical students or science master's programs as well as for researchers. It also serves common public interest by presenting new methods to improve the quality of life in general, with a better integration into society.

Utpal Chakraborty
Amit Banerjee
Jayanta Kumar Saha
Niloy Sarkar
Chinmay Chakraborty
January 2022

AI IN INDUSTRY 4.0

Chapter 1

Computer Vision–Based System for Automation and Industrial Applications

Huan Ngoc Le,[a] Ngoc Vuong Bao Tu,[b] and Narayan C. Debnath[c]

[a] *Mechanical and Mechatronics Department, Eastern International University, Nam Ky Khoi Nghia, Dinh Hoa, Thu Dau Mot, Binh Duong Province, Vietnam*
[b] *Senior Engineer, Street 10, Quang Trung Software Park, Tan Chanh Hiep Ward, District 12, Ho Chi Minh City, Vietnam*
[c] *School of Computing and Information Technology, Eastern International University, Nam Ky Khoi Nghia, Dinh Hoa, Thu Dau Mot, Binh Duong Province, Vietnam*
huan.le@eiu.edu.vn, tvbngoc@tma.com.vn, narayan.debnath@eiu.edu.vn

This chapter presents some new developments in mathematical frameworks, algorithms, applications, and case studies of computer vision–based systems that may prove useful in the manufacturing industry for reducing the defects in and improving the quality of products. In addition to the theory and design of the computer vision–based systems, this chapter includes detailed results and discussions of two important and practical applications of the automatic optical inspection (AOI) systems. One relates to AOI systems for detecting defects on electronic boards, and the other addresses an AOI system for detecting defects in the rubber keypads of a scanning machine. The new theory and prototype of the AOI systems together with proposed applications will have positive impacts in the manufacturing industries. Moreover, this research

Artificial Intelligence and the Fourth Industrial Revolution
Edited by Utpal Chakraborty, Amit Banerjee, Jayanta Kumar Saha, Niloy Sarkar, and Chinmay Chakraborty
Copyright © 2022 Jenny Stanford Publishing Pte. Ltd.
ISBN 978-981-4800-79-2 (Paperback), 978-1-003-15974-2 (eBook)
www.jennystanford.com

chapter, containing original research contributions in the field, may provide further encouragement in the design and development of new and improved computer vision–based systems that will provide significant benefits in terms of automation and encourage more industrial applications, especially while focusing on artificial intelligence and the fourth industrial revolution.

1.1 Introduction

Digital image processing technology was developed in 1920 and has many practical applications, such as image enhancement, artistic effects, medical visualization [1], industrial inspection, law enforcement, and human-computer interfaces. In the field of industrial inspection, automatic optical inspection (AOI) systems have been proposed and have served effectively in detecting certain types of surface-related defects. The system will collect image data of the object or product and send it to the central processor for analysis. The system is often used for testing involving measurement of assembly integrity, surface finish, and geometric dimensions of the product. The use of AOI increases productivity and improves product quality and is an option with many advantages over manual testing. With better camera manufacturing technology and automated equipment in combination with computer technology, including pattern recognition, image processing and artificial intelligence, big data, and machine learning, a test system can automatically run in real time and proves to be consistent, powerful, and reliable.

In most manufacturing industries, one of the main goals is quality assurance with zero percent defects in parts, accessories, and finished products. Accordingly, product defect testing becomes an important step in the manufacturing process. Besides, inspection is time consuming as it is typically carried out by human inspectors in most companies. Compared to machines, the inspection period of human inspectors is limited, the cost of the inspection process is high, and the performance of the human inspectors is often inadequate and inaccurate. A computer vision–based AOI system becomes an optimal solution for such problems. The AOI system can automatically detect surface-related defects and related issues.

The camera will take a picture of the product and send the image to the central processing unit for visual data analysis. Common tests include measurement of joint finishes, surface finishes, and geometric dimensions. Due to recent advances in computer vision technology, automated testing provides the opportunity to improve the quality of the final products and reduce cost in manufacturing industries. As a result, for manufacturers, a well-designed AOI system is a more appropriate choice compared to manual inspection.

The biggest challenge today is the production of affordable, compact, portable AOI systems that are easy to use for some of the most common testing purposes, such as missing components and wrong components, with acceptable accuracy.

To meet the needs of the enterprises for an efficient low-cost system that can identify the common types of defects in printed circuit board assemblies (PCBAs), such as missing or wrong components, a prototype of computer vision–based system was proposed. The objective is to develop theory and algorithms to automatically detect defects as quickly and accurately as possible, and to help achieve manufactured products of the highest possible quality. The study proposed a prototype of the computer vision-based system for automatically detecting defects on electronic boards, which includes a conveyor system and a capturing room. The system can run in two modes, manual and auto. A control algorithm was built to run on the Arduino microcontroller for speed control of a direct current (DC) motor and Visual Studio C# to analyze data of captured image and display the final result. To overcome the image distortion, the author proposed an effective correction method using image transformations defined by the points detected in the distorted image and the points in the undistorted image without using any model. For dealing with the rotation of input boards, a set of algorithms was developed, including speeded up robust features (SURFs) for detecting local features and a combination of k-nearest neighbors algorithm (k-NN) and Random sample consensus (RANSAC) for matching corresponding features, deleting outliers and generating rotation and translation parameters. As a result, the defective electronic board detection system can detect two common defects: missing electronic components and electronic components that are in the wrong positions. Moreover, users can create new data files for different board types and the files can also be edited, saved, and reused later. An AOI system consisting of a mobile image acquisition and processing unit has been developed for factory applications. The developed software package has been implemented on a mobile processing unit as an additional testing device at the factory. The mathematical theory and related algorithms for the system, along with the descriptions of the materials, methods, system components, and a case study, are presented.

In addition, a new AOI system application was developed that automatically detects rubber keypad defects and provides results similar to human classification experts. The system was designed to address all issues related to defects of rubber keypads before their manufacture, with the less time consumed. The study includes a quick calibration method without using any model for low-price lens camera. In this application, the object's surface is bent to coincide with the reference plane used in the calibration process. So, the size of all similar objects at any position in the image will be the same, even if they are not put in the same plane. This method showed a significant improvement in the result and allowed for speeding up the whole process to real-time processing. A mobile computer-based system with two operation modes was developed, and it showed improved performances with a wide range of keypads commonly used in the manufacturing industries. The program also permits users to easily add new types of keypads to the program database. The computing time for processing a patch of four keypads was significantly enhanced. The mathematical framework, along with the descriptions of the materials, methods, and algorithms, is provided. To prove the effectiveness of the developed algorithms, a detailed case study is presented where the algorithms have been tested directly with actual keypad devices at a local company.

1.2 Previous Research

Quality control and automatic visual inspection systems used in precision manufacturing have been extensively studied and discussed to facilitate the development of new production techniques.

The automatic visual inspection system is one of the most important automation tools used to reduce the workforce, reduce product costs, and increase output quality. As computer vision–based inspection has become one of the most important application areas, many studies and prototype systems have been proposed to test hardware conditions, software development, and related applications. Some of the studies focused primarily on specific products or techniques applied in automatic inspection are described in this section.

The authors in Ref. [2] used an AOI system to check and track the position state of test subjects in remote monitoring of a wafer or solar panel, and related optoelectronic devices. In Ref. [3], an online automatic optical inspection system using digital image processing technology has been developed to check the size distribution of many other coarse particles, such as sintered, limestone, dolomite, and serpentine. A computer vision–based measurement system in combination with a Programmable Logic Controller powered by the Object Linking and Embedding for Process Control technology was reported as a tool to determine the optimal economic setting for controlling and monitoring more than one process variables in industrial applications [4]. The AOI system has also been studied for the purpose of examining common defects on the edges of mirror components, such as scratches, impurities, and edge serration, that occur during the production of mirror elements [5]. In addition, a system for checking the microlens of the optical connector is presented in Ref. [6]. To detect the defects of rubber keypads, a vision-based high-speed system has been designed and developed [7].

To check the quality of input components before assembling them on printed circuit boards (PCBs) or to check the quality of the PCBA after it has been assembled, a lot of research and many practical applications have been carried out. In Ref. [8], a system was set up to automatically check the light-emitting diodes (LEDs) and be sure to select good samples for the PCBA line. Moreover, the system also calculates the yield rate of productivity during the inspection process. An AOI system was developed to diagnose the solder joint defects on PCBs assembled in surface-mount technology using a neural network in combination with genetic algorithms [9]. The authors in [10] conducted a comprehensive inspection of the weld a focus on three aspects: the recognition rate, which should be high; classification of defects, which should be detailed; and inspection speed, which should be fast. Accordingly, the authors have proposed a new detection and classification algorithm of the chip weld based on the color level and Boolean rules.

1.3 AOI System Application on a Scanning Machine

Today, factories must devote enormous human and financial resources for detecting defective rubber keypads. In this chapter, the authors have designed and developed a vision-based system for detecting a number of defects in rubber keypads. A new, easy-to-use method to improve image distortion has been developed on the basis of local transformations defined by the sets of points detected both in the undistorted image and in the distorted image without using any models. The proposed system can detect up to 14 different types of rubber keypads in about 1.8 sec. for a set of four keypads, even with a normal configured laptop and a cheap webcam.

1.3.1 Detecting Rubber Keypads Defects on a Scanning Machine

To ensure the quality of the keypad buttons and improve the productivity, checking for keyboard button errors is an important task in production. In this study, the authors designed a computer vision system to detect the defects in remote control keypad buttons. The defects in the keypad buttons are categorized as wrong position, wrong direction, wrong color, faulty font, missing buttons, etc., as shown in Fig. 1.1.

In Ref. [7], the authors presented automated testing using computer vision technology that can give manufacturers the opportunity to improve product quality and reduce production costs. The quality of the final products is greatly improved by reducing the defects at the component level. In some local companies (e.g., Datalogic at Saigon High Technology Center, Ho Chi Minh City, Vietnam), an employee used the eyes to check the quality of the products, needing

Figure 1.1   Defects in keypad buttons.

an average of 30 sec. to test a 52-button keypad. And the accuracy was just about 90%.

In the industry, managers always need systems that are as cheap, easy to use, and reliable as possible. Therefore, computer vision systems serving in companies tend to use wide-angle and low-price lens cameras in combination with commonly configured computers. Today, a wide-angle lens is widely used for many applications in photography, industrial equipment, and automation systems. Though there are many advantages of wide-angle images captured by a single camera, they always appear distorted at the periphery. Therefore, the image quality obtained is greatly reduced in applications that need high accuracy. To overcome this drawback, many authors have proposed a variety of lens distortion correction methods based on the use of new camera calibration schemes.

In a mobile robot tracking application, a distortion correction lens wide-angle camera was introduced in 2004 [11]. In 2015, in order to correct the fish-eye lens distortion, the authors in Ref. [12] proposed a nonlinear calibration method. The authors in Ref. [13] corrected and enhanced lens distortion by using local self-similarity. Other than these methods, to correct the nonlinear lens distortion, a new scheme without using a model was suggested [14]. Similar to the case in Refs. [14, 15], the authors presented a new calibration method to ignore nonlinear lens distortion. The difference from the traditional calibration method is that the above suggestions are used to deal with nonuniform distortions that conventional interpolation methods cannot completely solve.

To meet the requirements of a highly applicable system with a less time-consuming and high-precision algorithm, in this study, a new, fast distortion correction method that aims to use local transformations defined by the sets of points in both deformed and nondeformed images has been developed. After the analysis of the error distribution law with a checkerboard, an error compensation model using a simple interpolation method has been proposed.

For vision-based systems, identifying a fast, powerful automatic thresholding value is one of the most difficult issues. In the revolution of image processing techniques, the entropy method has been widely developed for automatic image segmentation. In Refs. [16, 17], the authors introduced and proposed for image segmentation a new entropy method called Tsallis entropy. Similarly, in Refs. [18–20], utilizing minimum cross entropy, several novel segmentation methods have been developed to find the appropriate threshold values. The maximum entropy that can be combined with spatial information and fuzzy entropy has also produced good results in automatic segmentation [21–25]. Usually, there are two basic ways to use the entropy method for automatic image segmentation, global and local. However, for these keypad applications, because the image size is large and the details are too small, the global approach is more appropriate.

Using the entropy method, the maximum entropy method, and a well-known method developed by Otsu [26], the authors compared the performances of several automatic segmentation schemes that have emerged as reliable schemes. Therefore, they have been studied in depth so that can be used and compared with the proposed algorithm.

The main objective of this study is to design and fabricate a simple, easy-to-use hardware system along with a software package to automatically detect keypad errors with the same high precision as human classification experts. The time consumed is specially considered so that the system can detect automatically in real time. Finally, the authors have completed the desired mobile system that makes it easier to collect and process images at the factory. The system consists of a storage area for keypads and a camera connected directly to the laptop. The entire system includes an error detection algorithm that has been tested directly with actual keypads at the company.

1.3.2 Theoretical Framework, Materials, and Methods

1.3.2.1 Mobile image processing unit

We developed a computer vision system consisting of an image acquisition unit (Fig. 1.2b) and a mobile computing unit mounted on a 15″ Notebook (Intel Core i7, 2.2 GHz, RAM 8 GB). A lighting system consists of circular white LEDs. The direct lighting LEDs are arranged in a rectangle around the camera, as shown in Fig. 1.2c. Figure 1.2d shows the image of the captured by using the image acquisition device. During this study, a big automation company making several types of keypads supported us by providing 14 different types of keypads. The analysis software is built in the C# language. As shown in Fig. 1.2a, a red line has been used to confirm that the keypad supplier is in a position to take a picture. When the system is started, the camera will take pictures continuously until the red line is displayed at the default position on the image. Then, the system will use this image as the input image. Figure 1.2e shows the 14 types of keypads, with the number of buttons varying from 29 to 52, used in this study.

1.3.2.2 Image calibration

To detect any defect on any keypad button, the simplest way is to find the corresponding button from the standard image and the one from the input image and use the matching method to determine the defect on the object to be searched. To use the matching method, the size of an object must be the same, irrespective of its position or direction in the image. Therefore, the authors have proposed a new calibration method that is easier to perform and gives higher accuracy than the traditional method.

Plane calibration and error After traditional calibration, objects near the camera will have a larger size in the image than the objects farther away from the camera. Therefore, for applications that need to measure the object size at every position on the same plane, even when the camera is not perpendicular to that position, we need a new calibration method. We have proposed a fast calibration method based on the nonmetric scheme. As in Refs. [27, 28], the closer

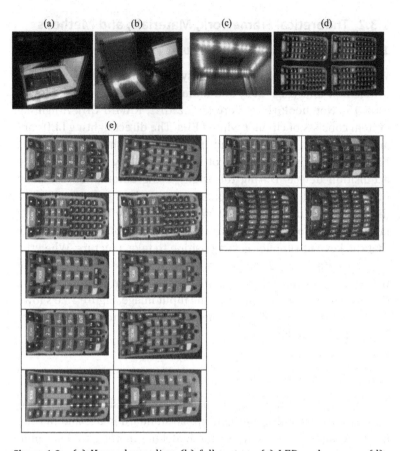

Figure 1.2 (a) Keypad supplier; (b) full system; (c) LED and camera; (d) sample image captured by the system; (e) 14 types of keypads used in this work.

the area to the image center, the lower is the radial distortion. In this study, we used a rectangular checkerboard 10 mm × 10 mm (56 × 56 pixels) to analyze the distribution law of errors. After that, we proposed a new calibration method in accordance with the requirements of the project. Accordingly, in order to use this method, it is supposed that the horizontal edges of the rectangle in the checkerboard are relatively parallel to the horizontal axis of the image frame.

The calibration process has the following steps:

(1) Capture the checkerboard image.
(2) Calculate the size of the centroid rectangle at the center of the image.
(3) Build a grid map based on the size of the centroid rectangle and the position of the image center.
(4) Take the shrinkage coefficients of the undistorted points or grid vertices versus the surrounding intersection points.
(5) Recover the input image based on the matrix of shrinkage coefficients.

Image processing process for obtaining intersection positions The algorithm flowchart applied to the image of the checkerboard to calculate the size of a centroid rectangle is shown in Fig. 1.3. The checkerboard image consists of alternating black and white squares. In this study, we used a 30 × 20 checkerboard with the squares measuring 1 cm². The checkerboard image will be taken once and saved in the database for future use. Every time the program is started, the program will reload this image and determine the calibration values. To detect intersections in the checkerboard, we first convert the color image to a gray-scale image and then a binary image using the Otsu scheme. The pixel intensity is now 0

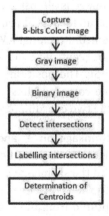

Figure 1.3 An algorithm using a flowchart is applied to calculate the size of a centroid rectangle.

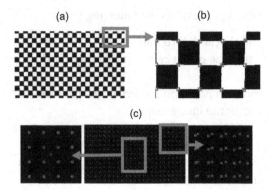

Figure 1.4 (a) At each point, check the area around it to find intersections. (b) The intersections are accurately marked in red in the checkerboard image. (c) The image of the dot-centroid position and grid vertices is added over the image of the intersection points.

(black) or 255 (white). We then start looking for the intersection points by checking all the pixels in the image. Figure 1.4a shows the intersections, accurately marked red in the checkerboard image. Figure 1.4b shows the position of the centroid point and the grid vertices added to the intersection points in the image.

Distribution of shrinkage In Ref. [29], a calibration error exists in both X and Y axes. As shown in Refs. [27, 28, 30], the calibration error in the center of the image is almost zero. So, we used the size of the centroid rectangle as an ideal grid cell. To create the grid, we first find the size of the centroid rectangle. We then create an ideal grid based on the size of the centroid square and the center of the image, as shown in Fig. 1.5c. The position difference between the intersection points and the corresponding grid vertices is used to analyze the shrinkage in the two coordinates X and Y.

The distribution law of shrinkage on X and Y axes In fact, the shrinkage is greater on the left edge and becomes smaller toward the center of the image. And the shrinkage increases gradually from the center of the image to the right edge. Through an analysis of a checkerboard image captured from the system, as shown in Figs. 1.5a and 1.5b, the shrinkage (in pixels) is estimated by the position difference between the vertex of the grid and the corresponding

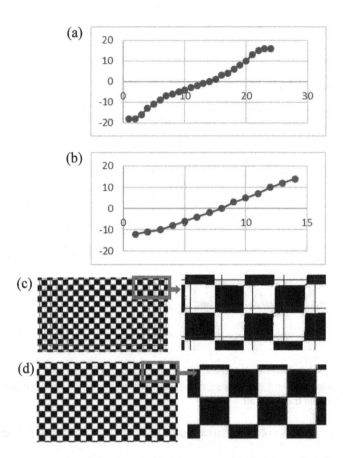

Figure 1.5 (a) Shrinkage distribution in X axis; (b) shrinkage distribution in Y axis; (c) the image of the checkerboard before calibration; (d) the image of the checkerboard after calibration.

intersection point. Images of the checkerboard before and after the calibration are shown in Figs. 1.5c and 1.5d.

Theory and method of generating the grid In the precorrected figure as shown in Fig. 1.5c, the coordinate values of the four vertices of each rectangle change to abscissas, as shown in Fig. 1.6a. We can find the center of the image as well as the four intersections closest to the center.

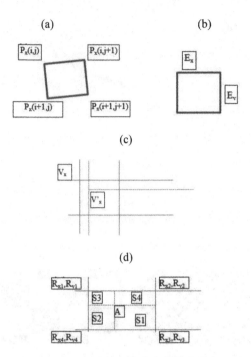

Figure 1.6 (a) Vertex-deviation in the input image; (b) grid cell; (c) vertex of the grid cell and the intersection point; (d) vertex of the grid cell and a point inside the grid cell.

The position of the central point is calculated as in Eqs. 1.1 and 1.2.

$$C_x(i, j) = [P_x(i, j) + P_x(i, j + 1) + P_x(i + 1, j + 1) + P_x(i + 1, j)]/4 \tag{1.1}$$

$$C_y(i, j) = [P_y(i, j) + P_y(i, j + 1) + P_y(i + 1, j + 1) + P_y(i + 1, j)]/4 \tag{1.2}$$

Then, using Eqs. 1.3 and 1.4, we can recalculate the size of the rectangle as shown in Fig. 1.6b.

$$E_x = [P_x(i, j + 1) + P_x(i + 1, j + 1) - P_x(i + 1, j) - P_x(i, j)]/2 \tag{1.3}$$

$$E_y = [P_y(i + 1, j) + P_y(i + 1, j + 1) - P_y(i, j + 1) - P_y(i, j)]/2 \tag{1.4}$$

On the basis of the size of the central rectangle, we can easily create a grid of rectangles of the same size as the central rectangle. Next, at each grid cell, the shrinkage coefficients in the X and Y directions are R_x and R_y, which can be achieved using Eqs. 1.5 and 1.6.

$$R_x = V_x / V_x' \tag{1.5}$$

$$R_y = V_y / V_y' \tag{1.6}$$

Here V is the distance of the vertices (distorted and undistorted) to the center of the image in Fig. 1.6c. From here on, for each vertex of the grid, we have the data of two shrinkage coefficients in the X and Y directions.

Construction of the error compensation model   To recover the input image, we create a zero-calibration image firstly. Then, at each pixel (A) belonging to a grid cell as shown in Fig. 1.6d, Eqs. 1.7 and 1.8 are used to estimate the $R_x A$ and $R_y A$ shrinkage coefficients at point A.

$$R_x A = (S1/S) * R_x 1 + (S2/S) * R_x 2 + (S3/S) * R_x 3$$
$$+ (S4/S) * R_x 4 \tag{1.7}$$

$$R_y A = (S1/S) * R_y 1 + (S2/S) * R_y 2 + (S3/S) * R_y 3$$
$$+ (S4/S) * R_y 4 \tag{1.8}$$

Here, $S = S1 + S2 + S3 + S4$

Then, from the shrinkage coefficients, we calculate the corrected position (AP) in the input image using the following equations:

$$XAP = XA / R_x A \tag{1.9}$$

$$YAP = YA / R_y A \tag{1.10}$$

Here, XAP and YAP are (X, Y) coordinates of the AP point and AP is the corrected position of A in the input image. However, since XAP and YAP are not integer values, in order to get the correct color value at the AP point, we need to again perform the interpolation using the following equation:

$$CA = CAP = (S1) * C1 + (S2) * C2 + (S3) * C3 + (S4) * C4$$
$$\tag{1.11}$$

Here, $S1, S2, S3$, and $S4$, as in Fig. 1.6d, are the areas of rectangles and $C1, C2, C3$, and $C4$ are the gray-scale values (0–255) of the four pixels around the AP point.

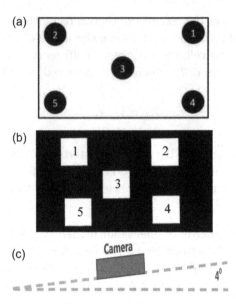

Figure 1.7 (a) Standard circle samples; (b) standard rectangular samples; (c) camera inclines to the base line with an angle of 4 degrees.

Verification of the compensation model We conducted experiments using circular samples to verify the accuracy of arc compensation. As shown in Fig. 1.7a, five circular samples 60 mm in diameter are located at five different positions on the calibration plate. After performing the calibration using the proposed method, we calculate the width and height before and after the compensation accordingly. Tables 1.1 and 1.2 show in detail the errors before and after correcting for both x and y directions, respectively.

After calibration, as shown in the data in Table 1.2, the dimensions of the circles are the same in any position. The experimental results show that the relative errors after compensation are reduced sharply. Another experience with five rectangles of the same size and placed in random positions in the camera view is shown in Fig. 1.7b. In Table 1.3, the shrinkage of the image has also been significantly improved.

It is difficult and takes time to point the camera directly at the surface of an object. Therefore, we verified the robustness of the

Table 1.1 The circle specification before calibration

Ex. no.	Size (pixels)	Y size (pixels)	X size (pixels)	Real size (cm)	Pixels/cm (Y axis)	Pixels/cm (X axis)	% error Y	% error X
1	20731	163	160	3	54.33	53.33	2.98	4.76
2	20496	162	159	3	54.00	53.00	3.57	5.36
3	21329	163	165	3	54.33	55.00	2.98	1.79
4	20066	159	159	3	53.00	53.00	5.36	5.36
5	20328	160	159	3	53.33	53.00	4.76	5.36

Table 1.2 The calibration result of circles

Ex. no.	Size (pixels)	Y size (pixels)	X size (pixels)	Real size (cm)	Pixels/cm (Y axis)	Pixels/cm (X axis)	% error Y	% error X
1	22415	168	167	3	56.00	55.67	0.00	0.60
2	22130	167	167	3	55.67	55.67	0.60	0.60
3	22424	168	168	3	56.00	56.00	0.00	0.00
4	22304	168	168	3	56.00	56.00	0.00	0.00
5	22407	169	167	3	56.33	55.67	0.60	0.60

proposed method by tilting the camera along the horizontal axis of the object surface up to 4 degrees, as shown in Fig. 1.7c.

As we saw in Fig. 1.8a, the horizontal stripes on the left are larger than those on the right in the input image. After applying the proposed calibration method, we get the result shown in Fig. 1.8b. The stripes on the left and the right are of the same size. Similarly, the algorithm still works well while the experimental sample is set up with vertical stripes, as shown in the Figs. 1.8c and 1.8d.

1.3.2.3 Image segmentation

During the process, there are two times we need to segment the image into two regions, objects and background. First, we need to separate the keypad from the supplier. In this step, we used an adaptive segmentation using the Otsu scheme [26] for both the reference image and the input image. Then, we need to get the white

Table 1.3 The calibration result of rectangles

	Real size		Measurement before calibration		Error before calibration	Measurement after calibration		Error after calibration
	Area (cm^2)	Edge length (cm)	Area (pixels)	Area (cm)	Area (%)	Area (pixels)	Area (cm)	Area (%)
Ob1	16	4	84225	15.38	3.87	87187	15.92	0.49
Ob2	16	4	83437	15.24	4.77	86591	15.81	1.17
Ob3	16	4	86757	15.84	0.98	88063	16.08	0.51
Ob4	16	4	83218	15.20	5.02	86468	15.79	1.31
Ob5	16	4	85066	15.53	2.91	87432	15.96	0.21

characters out of the buttons. As seen in Fig. 1.1, the background color of the buttons is different. Therefore, after the rectangles are positioned on the buttons in the reference image, local segmentation is performed for each button. In this study, we compared the entropy segmentation methods [17–27] and a proposed method as in the following formula to identify the most suitable scheme.

$$I(x, y) = \begin{cases} 255 & \text{if} \\ 0 & \text{otherelse} \end{cases} g(x, y) > \frac{\max G * 2 + \min G}{3} \right\}, \quad (1.12)$$

where $\max G$ and $\min G$ are the maximum and minimum gray levels in the local area, respectively. Finally, on the basis of the stability and accuracy of the results, the proposed method was chosen as a local segmentation scheme in the reference image.

In the input image, we use the correlation method to check whether the button exactly matches the button in the reference image. To take the white characters out of the buttons from the input image, an adaptive segmentation method is proposed and implemented for each button, as shown below.

Step 1: Estimate the total number of while pixels in rectangle holding reference button (T_{pixel}).

Step 2: In rectangle holding corresponding button in the input image, do thresholding with the threshold value of Thres_value $= 255 - 1$.

Step 3: Check whether the number of white pixels is equal to or greater than T_{pixel}.

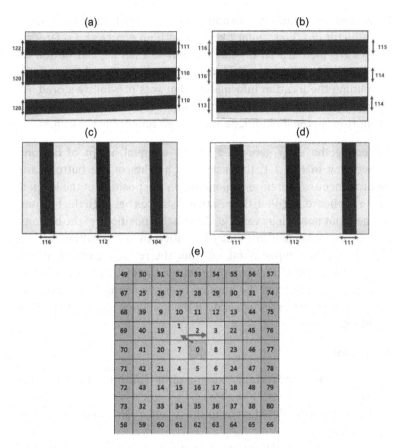

Figure 1.8 (a) Horizontal stripes before calibration; (b) horizontal stripes after calibration; (c) vertical stripes before calibration; (d) vertical stripes after calibration; (e) scanning scheme.

Step 4: Stop the process if Step 3 is correct or Thres_value < 0. Otherwise, go back to Step 2 with Thres_value = 1.

1.3.2.4 Automatic defect detection algorithm

In this study, we seek a solution that can detect defects automatically with the highest accuracy and in the shortest processing time possible.

Defect detection method   To find defects on each keypad button, we use a picture of the standard keypad with no errors as a reference image. We then search for the location of each keypad button in the keypad reference image surrounded by rectangles in a file containing the location information that was previously stored. At each keypad button, we use the matching method to check whether the buttons in the input image and the reference image match.

Correcting the error generated by the unequal height of buttons As we saw in Fig. 1.1, the sizes and heights of the buttons are heterogeneous. Therefore, depending on the position of the keypad in the keyboard supplier, the relative distances between the buttons on the input image may vary. So, the relative positions of the buttons on the input keypad are shifted relative to the positions of the buttons on the keypad obtained from the reference image. When the keypad is located far from the center, the distance between the buttons increases from the buttons near the center to the buttons far from the center, so we proposed the following method to fix this problem.

Definitions

(x,y): Position of the rectangle surrounding a button in the reference image.

(x_e, y_e): Accumulated error in the X and Y directions. Initialize: $x_e = 0, y_e = 0$.

(x_r, y_r): Used to determine the new position of the rectangle in the detected image on the basis of the position of the rectangle in the reference image. (x_r, y_r) are estimated on the basis of the rule shown in Fig. 1.8e. Initialize: $x_r = 0, y_r = 0$.

$(x + x_r + x_e, y + y_r + y_e)$: Position of the rectangle in the detected image.

Scanning process

(1) Go to the first rectangular position in the reference image, (x,y).
(2) Do thresholding and check the correlation inside the two rectangles of size (w,h) at (x,y) in the reference image and $(x + x_r + x_e, y + y_r + y_e)$ in the detected image. The program compares the binary values of each pair of corresponding pixels

in both windows. If the number of matchings in both black and white is greater than the correlation coefficient that can be changed by the user, this means a good button has been found. The correlation coefficient being used is 92%.

(3) If $x_r = 4$ and $y_r = 4$ and no match was found, a defect has been detected. Go to Step 2 with the next rectangular position.

Else, if a match is found, $x_e = x_e + x_r/2$ and $y_e = y_e + y_r/2$. Go to Step 2 with the next rectangular position.

Detecting defects in mode 1: A keypad put in a quadrant By understanding the actual requirements in the production process, we decided to develop two operating modes that have mutual effects. In mode 1, the keypad supplier is divided into four equal quadrants. The detected keypads must be located in the quadrants of the keypad supplier. We proposed a defective detection algorithm with the flowchart of the sequence as shown in Fig. 1.9. In this mode, the technician spends more time placing the keypad correctly into the quadrants of the supplier.

As shown in the diagram, first the camera takes a picture and transfers the image to a step that calibrates the input image using the method proposed above. Then, at each quadrant of the image, we look for the locations of the reference (Ref.) and SCAN buttons. These are two colored buttons so the technician can easily recognize them if they are missing. If neither button is found (the keypad does not exist), the program automatically switches to the next quadrant. Similarly, if one of the two buttons is not found, we assume there is an error in this keypad and the program automatically switches to the next quadrant. In the quadrant where there is a keypad under review, the program determines the exact position of the Ref. and SCAN buttons. Then, using the relationship between the Ref. and SCAN buttons and the X axis, it is easy to move and rotate the keypad horizontally in the conventional orientation in the same direction as the reference keypad. Finally, we examine the defection of each button by comparing each pixel in the button in both the input image and the reference image one by one.

Detecting defects in mode 2 Up to four keypads can be put anywhere in the keypad supplier. In mode 1, the operating time of

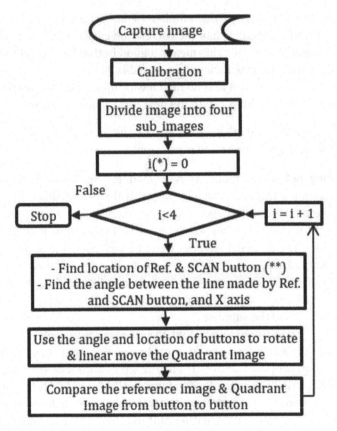

Figure 1.9 Flowchart of the automatic defect detection in mode 1. (*): The *i* indicates order of the image. (**): In this project, Ref. and SCAN buttons were defined as two fixed reference buttons defined by the user.

the workers was not optimal because they had to put the keypads in quarters, so we developed this control mode, mode 2. We can put the keypads anywhere in the supplier in this mode. We proposed the sequence diagram used to identify the errors as shown in Fig. 1.10. In Fig. 1.11a, the keypad supplier of our system can accommodate up to four keypads.

After taking a picture, first the system performs image calibration (Fig. 1.11b). The threshold value is then determined by the Otsu method to extract the keypad from the background (Fig. 1.11c). Then, the closing morphology is performed once with a 5×5

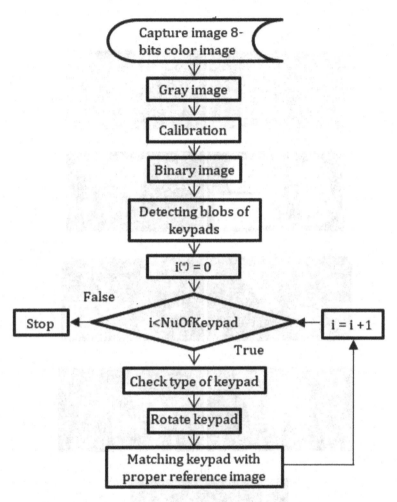

Figure 1.10 Flowchart of the automatic defect detection. (*): The *i* indicates order of the image.

rectangular mask to fill the small holes in the keypads. Next, to count the major blobs that are the keypads, we use the labeling algorithm. To handle a single keypad in an image, we proceed to separate the large blobs into separate images. In each individual image, we fill in the small black spots inside the big blobs and replace the pixel values of the white spots with the pixel values of the input image as seen in Fig. 1.11d–f. Then, we move the keypad in the middle of the image

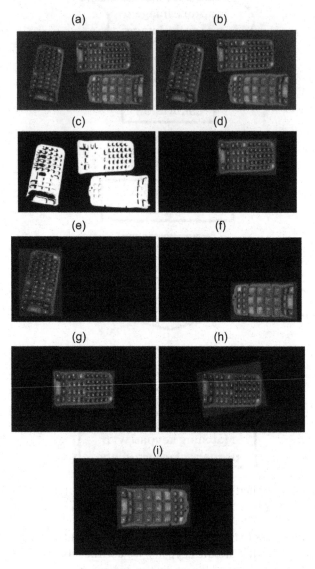

Figure 1.11 (a) Input image; (b) resulting image after calibration; (c) binary image and detecting the big blobs; (d–f) resulting image after detecting the type of keypads; (g–i) resulting image after the keypads are moved and rotated horizontally.

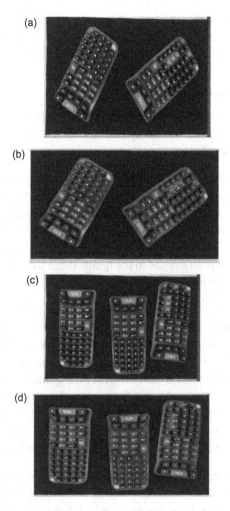

Figure 1.12 (a) Four input keypads; (b) three input keypads; (c) resulting image of (a) after applying the mode 1 algorithm; (d) resulting image of (c) after applying the mode 1 algorithm.

and try to rotate the image so it matches the orientation and position of the reference image. To do this, the SCAN button is searched for inside each keypad. We move to the next image if no SCAN button exists. If there is a SCAN button, we calculate the Pre_Angle angle created by the SCAN button and the center of the image relative to the X axis. The Pre_Angle angle is then used to rotate the keypad so that the keypad is placed horizontally (Figs. 1.11g–i). Finally, all the buttons of the keypad are checked for defects by comparing each button in the input image with that in the reference image.

1.3.2.5 Results and discussion

In both modes, all types of errors can be detected with equal success. Accordingly, the users can easily add new types of keypads into the program database by editing and saving the necessary features.

Mode 1   In Fig. 1.12a, four keypads are located in the four quarters in the keypad supplier. Three keypads have defects identified and highlighted by the red rectangles that overlap the buttons as shown in Fig. 1.12b. Similarly, Fig. 1.12d displays the results after checking the defects of the keypads shown in Fig. 1.12c. The algorithm works fine even if there is no keypad in the quadrant or the keypad is placed in the opposite direction to the other keypads.

Mode 2   This is a more complete mode than mode 1. Technicians need less time to locate the keypads above the supplier than in mode 1. Therefore, the software needs to implement many more operations and the program also takes more time than the previous mode. However, the fact is that the processing time of both modes is not too different. As in the previous case, the proposed algorithm works well when the keypad is placed in any position in any direction. Figures 1.13b and 1.13d present the results of the input images as shown in the Figs. 1.13a and 1.13c, respectively.

Economic efficiency of applying the research in manufacturing   During the implementation of the project, at the surveyed companies, all workers were responsible for checking the defects of the rubber keypads with their eyes. According to the general statistics, for a 52-button keypad, a worker took about 30 sec. to detect the errors on a keypad. However, in the end, about 10% of the errors were missed. The final software version of this study showed that the computation time for processing a batch of four keypads is approximately 1.8 sec. and 2.3 sec. for mode 1 and mode 2, respectively, as in Table 1.4. The system has been used for physical checks hundreds of times and shows very accurate results. This system will be applied practically to the industrial companies associated with the project.

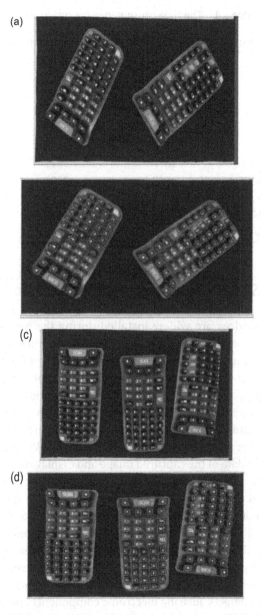

Figure 1.13 (a) Two input keypads; (b) resulting image of (a) after applying the mode 2 algorithm; (c) three input keypads; (d) resulting image of (c) after applying the mode 2 algorithm.

Table 1.4 The time consumed in the two modes

Sample	Time (ms) mode 1	Time (ms) mode 2
1	1798	2280
2	1880	2410
3	1801	2365
4	1666	2131
5	1902	2479
6	1934	2472
7	1877	2301
8	1788	2258
9	1803	2210
10	1872	2410
Average	1832	2332

1.4 AOI System Application on Electronic Boards

Today, AOI devices are widely installed during the design, layout, fabrication, assembly, and testing of production lines to improve productivity and reduce production costs. This study proposed a prototype of a computer vision–based system for automatically detecting defects on electronic boards, which includes a conveyor system and a capturing room. The system can run in two modes, manual and auto. A control algorithm was built to run on the Arduino microcontroller for speed control of a DC motor and Visual Studio C# to analyze the data of the captured image and display the final result. As discussed in Section 1.3.2.2, in this study, we took advantage of the proposed method to correct image distortion. For dealing with the rotation of the input boards, we developed a set of algorithms, including SURF for detecting local features and a combination of k-NN and RANSAC for matching corresponding features, deleting outliers, and generating rotation and translation parameters. As a result, the defective electronic board detection system can detect two common defects, missing electronic components and electronic components in the wrong position. Moreover, the users can create new data files for different board types and they also can be edited, saved, and reused later.

1.4.1 Detecting Defects on Electronic Boards

Today, with the rapid growth of smart electronics as well as home appliances, the number of factories manufacturing PCBA devices is extremely large. The quality control of PCBAs is extremely important. This is a time-consuming and costly investment. In many companies, PCBA's quality control (number of components, weld quality, properly installed components, etc.) is mainly done by human inspectors. However, in the 4.0 era, people are increasingly separated from the manufacturing process if automatic devices can do it faster and more efficiently. The AOI system has been long developed and is used as an effective device in detecting product surface defects, such as assembly integrity, surface finish, and geometric dimensions.

In Ref. [31], the authors applied a Fourier descriptor–based image alignment algorithm to an AOI system to check the quality of the PCBA in a real-time environment. In addition, in 2014, a team of authors used the color level and Boolean rules developed in a study [32] to develop algorithms to improve the overall performance of solder joints in terms of three aspects: the recognition rate, which should be high; classification of defects, which should be detailed, and inspection speed, which should be fast. In particular, the Boolean rule is used to detect eight common types of solder joints: acceptable, pseudo, no solder, lacked solder, excess solder, shifted, tombstone, and miss component. Similarly, in Ref. [33], the authors developed a neural network in combination with genetic algorithms to detect PCB defects. This AOI system can classify six common types of solder joints in practice.

In all of the above research, the authors conducted detailed studies on the welding of the component parts using accurate mechanical systems equipped with expensive closed-circuit-device cameras and light illumination systems for help in the imaging process. The devices can identify up to seven different types of solder joints in PCBAs, and the processing time is higher than 15 sec. for a PCBA.

An image distortion correction algorithm is always required for image processing systems that need extremely accurate results. In this study, we take advantage of the distortion correction method as detailed in Section 1.3.2.2 to calibrate the input image before processing.

It is costly and time consuming to set up the system so that the circuit board position is fixed and the camera moves to the desired locations, as in expensive AOIs. We can place the PCBA freely on the conveyor belt, and the detection algorithm will rotate the board to the desired position using the software. For dealing with the input image registration, we developed a set of algorithms, including SURF [10] for detecting local features and a combination of k-NN [34] and RANSAC [35] for matching corresponding features, deleting outliers, and generating rotation and translation parameters.

To match the actual requirements of the enterprise for an efficient, low-cost system that can identify the common types of defect, such as missing or wrong components, this paper proposed a low-cost AOI system. The main objective of this study is to develop algorithms to automatically detect PCBA defects as quickly and accurately as possible with the results close to those of human inspectors. Therefore, we have developed a mobile image acquisition and processing system for on-site applications. The algorithms developed can complement a simpler, easier-to-use, less expensive hardware system to meet the high demands of most companies that have difficulties accessing existing expensive AOI systems on the market.

1.4.2 Theoretical Framework, Materials, and Methods

1.4.2.1 Mobile image processing unit

We developed a computer vision system consisting of an image acquisition unit, a conveyor, and a mobile computing unit mounted on a 15″ Notebook (Intel Core i7, 2.2 GHz, RAM 8 Gb), as shown in Fig. 1.14a. The illumination system was composed of a white LED rectangular-type direct lighting device. A web camera was placed at the center of the rectangular-type LED lighting at the top site of the image acquisition unit. PCBAs were supported by a big automation company making several types of PCBAs. The analysis software was made with C# and sample PCBA images (Fig. 1.14b) taken directly from the conveyor. A photo sensor was used for checking the

(a)   (b)

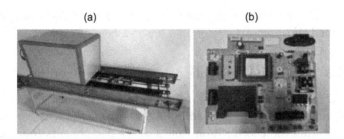

Figure 1.14   (a) A full prototype of a defective detection system; (b) a PCBA image captured by the system.

position of PCBA relative to the camera. The speed of the conveyor is changeable and controlled by an Arduino Mega 2560 controller.

1.4.2.2 Image calibration

Plane calibration   In this system, the camera is always normal and at a constant distance from the conveyor. So, we proposed a quick calibration method based on a nonmetric scheme. As in Refs. [28, 36], the nearer the area around the image center, the lower is the radial distortion. In this research, a rectangular checkerboard 10 mm × 10 mm (56 × 56 pixels) in size was used to analyze the distribution law of the error. To use the suggested method, it is supposed that the horizontal edges of the rectangles in the checkerboard are parallel relative to the horizontal axis of the image frame.

The calibration process is as follows:

(1) Capture the checkerboard image.
(2) Calculate the size of the centroid rectangle at the center of the image.
(3) Build a grid map based on the size of the centroid rectangle and the position of the image center.
(4) Take the shrinkage coefficients of the undistorted points or grid vertices versus the surrounding intersection points.
(5) Recover the input image based on the matrix of shrinkage coefficients.

The calibration scheme used in this study is the same as described above.

1.4.2.3 Editing and automatic defect detection algorithm

Before testing PCBA's quality, we have to edit the test mode for each component on a standard PCBA. In general, the user can get a file storing the coordination values of the components from the PCB maker. From there, the program can locate the components exactly and the users edit the test mode for each component after that. However, most PCBAs that need to be tested do not have the coordination data file. Therefore, in this paper, the authors have proposed a manual method and a semiautomatic method to locate the components on each PCBA image.

Manual editing   In Fig. 1.15a, the Hardware Setting tab gives the user the ability to configure the serial port, DC servo parameters, and the motor's speed. It also allows the user to change the Kp, Ki, and Kd coefficients for PID control, change the exposure of the camera, and control the lighting to ensure a quality image.

To edit the component's configuration manually, first, we capture a standard image (image without errors) of the PCBA and go to the Editing tab (Fig. 1.15a). On the user interface is a button for generating rectangles of variable sizes. We can click and drag the rectangle to the desired position of the component and begin to

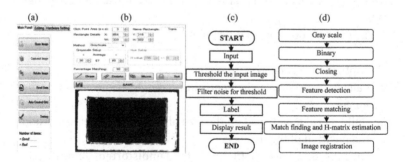

Figure 1.15   (a) The main tab of the user interface, (b) setting the area for each component, (c) a flowchart of automatic component registration, and (d) a flowchart of the automatic image registration method.

configure the values for detecting the component, such as the threshold value, the matching percentage, and the type of used color (gray scale or hue), before saving the component for later use (Fig. 1.15b). At each rectangle, the user can adjust a proper threshold value by seeing the segmentation result directly below the Save button. The functions to find the average value and hue (H) are shown in Eqs. 1.13 and 1.14, respectively. Then we continue configuring sequentially for all necessary components. Finally, an information file including the location and the test mode of all components on the PCBA will be saved and used for actual inspection.

$$\text{Average value} = \frac{\text{Red} + \text{Green} + \text{Blue}}{3} \quad (1.13)$$

$$R' = \frac{\text{Red}}{255}; G' = \frac{\text{Green}}{255}; B' = \frac{\text{Blue}}{255}$$

$$C_{\max} = \max(R', G', B')$$

$$C_{\min} = \min(R', G', B')$$

$$\Delta = C_{\max} - C_{\min}$$

$$H = \begin{cases} 0°, \Delta = 0 \\ 60° \times \left(\frac{G'-B'}{\Delta}\right), C_{\max} = R' \\ 60° \times \left(\frac{B'-R'}{\Delta} + 2\right), C_{\max} = G' \\ 60° \times \left(\frac{R'-G'}{\Delta} + 4\right), C_{\max} = B' \end{cases} \quad (1.14)$$

Here, Red is the intensity of the image in the red color frame, green is the intensity of the image in green color frame. And, Blue is the intensity of the image in the blue color frame.

Semiautomatic component registration Manual editing takes a lot of time when the number of components on a PCBA is large. Therefore, we, the authors, suggested a semiautomatic editing mode for finding the location and displaying the rectangle surrounding the existing components on the PCBA. We proposed an automatic object registration method with the flowchart as shown in Fig. 1.15c. In the flowchart, we first select a threshold value suitable for each

PCBA and segment the image to separate the components of the background. The image after thresholding is processed successively by dilation and erosion morphological algorithms (two erosions and two dilations). Then, we continue labeling the remaining blobs. We draw a rectangle around each blob. Finally, we get the locations and sizes of all the rectangles and superimpose these rectangles over the original image for re-editing manually (done by the users).

Image registration   With an input image placed randomly, there will be two kinds of rotations compared to the reference image. First, the input image can rotate 180 degrees from the reference image. Then, even if it is placed in the same direction, the input image can be rotated a little compared to the reference image. Therefore, before proceeding to find the local errors on the board, we must rotate the input image so it is in the same direction as and parallel to the reference image. To solve this problem, we first convert the color image to a gray-scale image. Then, we use the Otsu method to find an adaptive threshold value for the segmentation step. We use a closing morphological scheme for keeping the general shape of the object, and from that we can find features in both input and reference images using SURF. Then, we use the RANSAC algorithm to filter matches from the previous step. From these matches, we find the homography matrix (H-matrix) and use it to rotate and translate the test image to match with the reference image. Figure 1.15d shows the flow chart of automatic image registration.

Matching operation   In each rectangle, we use the saved data, including the threshold value range and the type of color, to run the thresholding operation for both the reference image and the captured image. Then, we compare the value of each pixel on both images. If the value is similar (both black and white), the counter will increase by 1. After that, we calculate the real matching percentage (RMP) by using Eq. 1.15.

$$\text{Real matching percentage} = \frac{\text{The value of the counter}}{\text{The area of the rectangle}} \quad (1.15)$$

The final step is the comparison of the RMP and the default matching percentage (DMP) of this rectangle. If the RMP is greater than or equal to the DMP, this rectangle will be displayed in green color. And

(a)   (b)

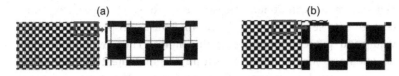

Figure 1.16 (a) The image of the checkerboard before calibration; (b) the image of the checkerboard after calibration.

if the RMP is less than the DMP, this rectangle will be displayed in red color.

1.4.2.4 Result and discussion

Image calibration Figures 1.16a and 1.16b show the images of the checkerboard before and after calibration, respectively. In this, if the object is put in the same plane on the checkerboard while calibration, the image distortion will be corrected, as in traditional calibration. By contrast, using the checkerboard's surface as a standard, the object's surface will be bent to coincide with the reference plane. This helps make a highly accurate match.

Manual editing When we need to test a new PCBA, first calibration is done. Then, on the basis of a standard board, we determine the positions of the components to be tested and the necessary parameters, such as gray color or hue and percentage matching. The result after configuring all the components on a sample board is shown in Fig. 1.17a.

(a)   (b)

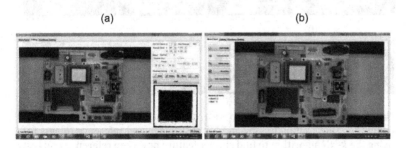

Figure 1.17 (a) The result after configuring all the components on a sample board; (b) superimposing rectangles onto the PCBA after automatic component registration.

Semiautomatic component registration To start registering the components semiautomatically, we find the appropriate threshold value and proceed to segment the image. Then, we implement an opening morphological algorithm to eliminate the noise that appears in the binary image. On implementing the steps as shown in the flowchart in Fig. 1.15c, the result is that all the components have rectangles around them (Fig. 1.17b).

After automatically finding PCBA components, it is normal to have components marked incorrect or rectangles that are too small or too large for the components. Therefore, we need to check for these errors and edit again manually. For PCBAs with more than 100 components, implementing a semiautomatic method for positioning components saves a lot of time compared to doing it manually.

Image registration The image registration process is implemented as shown in the flowchart in Fig. 1.15d. The input image, Fig. 1.18a, is converted to a gray-scale image, and the result is shown in Fig. 1.18b.

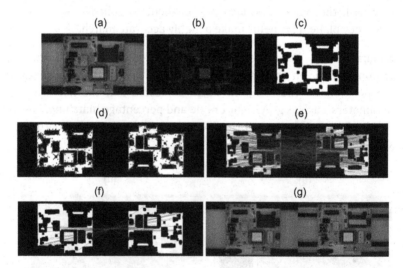

Figure 1.18 (a) Input image, (b) gray-scale image, (c) binary image, (d) using the SURF finding feature in both test image and reference image, (e) using the *k*-NN algorithm to find similar feature points on reference and test images, (f) using the RANSAC algorithm to filter matches from the previous step, and (g) using the Homography matrix obtained in the previous step to rotate and translate the test image to match with the reference image.

We then segment the gray-scale image using the Otsu scheme and continue implementing a closing morphological algorithm to keep the general shape of the object intact (Fig. 1.18c). Figure 1.18d shows the result of the features detected in both the image and the reference image using SURF. Then, after using the k-NN algorithm to locate similar feature points on the reference and test images, we get the result as in Fig. 1.18e. In Fig. 1.18f, we filter matches from the previous step using the RANSAC algorithm. Finally, using the homography matrix obtained in the previous step, we rotate and translate the test image to match with the reference image (Fig. 1.18g).

The final result Once we have the configuration data file of the standard PCBA, we run the proposed algorithm to determine the error of the PCBA running on the conveyor. The program can check exactly two common mistakes, a lack of components and the wrong type of components. Figures 1.19c and 1.19d are the results after examining the defective PCBA boards in Figs. 1.19a and 1.19b, respectively.

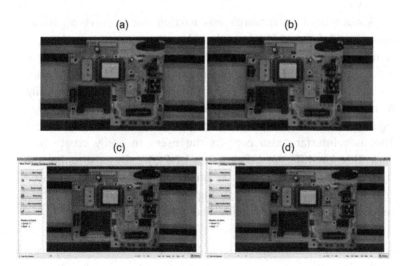

Figure 1.19 (a, b) Two test images; (c) is the result of (a); (d) is the result of (b).

1.5 Conclusions

Computer vision systems in the industry in general and to support the detection of defects in particular have been widely used, especially in the Industry 4.0 era. In this study, we have built a complete solution to solve all the problems related to defect detection on rubber keyboards in less amount of time before they are assembled into the final product. In the process of implementing the project, we also proposed a method for quick calibration without using any model for a wide-angle and low-cost lens camera. In particular, the object surface will be bent to coincide with the reference plane used during the calibration process. Therefore, after calibration, the dimensions of all the same objects in any position in the image will be the same even if they were not placed in the same plane. This calibration method gives important results in terms of accuracy and processing time. This is a decisive factor to accelerate the entire process in real time. The developed mobile-based system can work in two different modes, and both have shown significant results with 14 different keypad types. Users are allowed to easily add features of similarly sized keypads into the program database.

Besides, another research was carried out to solve all issues related to defects of PCBAs with a cheap system compared to the commercial, expensive AOI systems. The program works well even if the input board is rotated in comparison to the reference board because it searches for and rotates the input board automatically. A mobile computer-based system was developed, and it showed significant results, using about 9 sec. per PCBA with over 50 objects. The user interface also permits the users to easily create new data files for different board types of PCBAs by positioning the component manually and semiautomatically. Although the proposed algorithms have worked well, improving the system to further reduce processing time is highly desirable.

References

1. Chakraborty, C. (2017). Chronic wound image analysis by particle swarm optimization technique for tele-wound network. *Wireless Pers. Commun.*, **96**(3):3655–3671.

2. Lin, C. S., Tzeng, G. A., Cheng, C. T., Lay, Y. L., Tien, C. L. (2014). An automatic optical inspection system for the detection of three parallel lines in solar panel end face. *Optik*, **125**:688–693.

3. Liao, Y. S., Tarng, Y. S. (2009). On-line automatic optical inspection system for coarse particle size distribution. *Powder Technol.*, **189**:508–513.

4. Aydogmus, O., Talu, M. F. (2012). A vision-based measurement installation for programmable logic controllers. *Measurement*, **45**:1098–1104.

5. Lin, C. S., Chen, Y. H., Chiang, T. F., Lee, J. W., Lin, W. C., Lin, Y. D. (2018). An automatic inspection system for the coating quality of the edge of mirror elements. *Optik*, **152**:73–83.

6. Lin, C. S., Loh, G. H., Tien, C. L., Lin, T. C., Chiou, Y. C. (2013). Automatic optical inspection system for the micro-lens of optical connector with fuzzy ratio analysis. *Optik*, **124**:3085–3090.

7. Gang, D. C., Han, S. I., Lee, B. G., Lee, J. J. (2007). Keypad inspection system of cellular phone. In *Proceedings of Computer Graphics, Imaging and Visualisation (CGIV 2007)*, Bangkok, pp. 93–96.

8. Fung, R. F., Yang, C. Y., Lai, C. T. (2011). Graphic supervisory control of an automatic optical inspection for LED properties. *Measurement*, **44**:1349–1360.

9. Hao, W., Xianmin, Z., Yongcong, K., Gaofei, O., Hongwei, X. (2013). Solder joint inspection based on neural network combined with genetic algorithm. *Optik*, **124**:4110–4116.

10. Wu, F., Zhang, X. (2014). An inspection and classification method for chip solder joints using color grads and Boolean rules. *Rob. Comput. Integr. Manuf.*, **30**:517–526.

11. Klančar, G., Kristan, M., Karba, R. (2004). Wide-angle camera distortions and non-uniform illumination in mobile robot tracking. *Rob. Auton. Syst.*, **46**:125–133.

12. Tiwari, U., Mani, U., Paul, S. (2015). Non-linear method used for distortion correction of fish-eye lens: comparative analysis of different mapping functions. In *Proceedings of International Conference on Man and Machine Interfacing (MAMI)*, Bhubaneswar, pp. 1–5.

13. Kim, D., Park, J., Jung, J., Kim, T., Paik, J. (2014). Lens distortion correction and enhancement based on local self-similarity for high-quality consumer imaging systems. *IEEE Trans. Consum. Electron.*, **60**:18–22.

14. Ricolfe-Viala, C., José, A., Salmerón, S. (2010). Correcting non-linear lens distortion in cameras without using a model. *Opt. Laser Technol.*, **42**:628–639.

15. Aritan, S. (2010). Efficiency of non-linear lens distortion models in biomechanical analysis of human movement. *Measurement*, **43**:739–746.

16. Portes de Albuquerquea, M., Esquef, I. A., Gesualdi Mello, A. R. (2004). Image thresholding using Tsallis entropy. *Pattern Recognit. Lett.*, **25**:1059–1065.

17. Lin, Q., Ou, C. (2012). Tsallis entropy and the long-range correlation in image thresholding. *Signal Proc.*, **92**:2931–2939.

18. Nie, F., Gao, C., Guo, Y., Gan, M. (2011). Two-dimensional minimum local cross-entropy thresholding based on co-occurrence matrix. *Comput. Electr. Eng.*, **37**:757–767.

19. Tang, K., Yuan, X., Sun, T., Yang, J., Gao, S. (2011). An improved scheme for minimum cross entropy threshold selection based on genetic algorithm. *Knowledge Based Syst.*, **24**:1131–1138.

20. Yin, P. Y. (2007). Multilevel minimum cross entropy threshold selection based on particle swarm optimization. *Appl. Math. Comput.*, **184**:503–513.

21. Brink, A. D. (1996). Using spatial information as an aid to maximum entropy image threshold selection. *Pattern Recognit. Lett.*, **17**:29–36.

22. Lan, J., Zeng, Y. (2013). Multi-threshold image segmentation using maximum fuzzy entropy based on a new 2D histogram. *Optik*, **124**:3756–3760.

23. Tang, Y., Mu, W., Zhang, Y., Zhang, X. (2011). A fast recursive algorithm based on fuzzy 2-partition entropy approach for threshold selection. *Neurocomputing*, **74**:3072–3078.

24. Yu, H. Y., Zhi, X. B., Fan, J. L. (2015). Image segmentation based on weak fuzzy partition entropy. *Neurocomputing*, **168**:994–1010.

25. Zhou, J., Chen, L., Chen, C. L. P., Zhang, Y., Li, H. X. (2016). Fuzzy clustering with the entropy of attribute weights. *Neurocomputing*, **198**:125–134.

26. Otsu, N. (1979). A threshold selection method from grey level histogram. *IEEE Trans. Syst. Man Cybern.*, **9**:62–66.

27. Ricolfe-Viala, C., Sanchez-Salmeron, A. J. (2011). Using the camera pin-hole model restrictions to calibrate the lens distortion model. *Opt. Laser Technol.*, **43**:996–1005.

28. Sun, Q., Hou, Y., Chen, J. (2015). Lens distortion correction for improving measurement accuracy of digital image correlation. *Optik*, **126**:3153–3157.

29. Malis, E., Chaumette, F., Boudet, S. (1998). 2D 1/2 visual servoing stability analysis with respect to camera calibration errors. *IEEE/RSJ*, **2**:691–697.

30. Gao, D. K., Wang, Y. Q., Zhou, C. L., Xu, Z. P. (2012). The 2D calibration error analysis and compensation in the visual measurement. In *Proceedings of 4th International Conference on Intelligent Human-Machine Systems and Cybernetics (IHMSC)*, pp. 238–241.

31. Chen, C. S., Yeh, C. W., Yin, P. Y. (2009). A novel Fourier descriptor based image alignment algorithm for automatic optical inspection. *J. Visual Commun. Image Represent.*, **20**:178–189.

32. Wu, F., Zhang, X. (2014). An inspection and classification method for chip solder joints using color grads and Boolean rules. *Rob. Comput. Integr. Manuf.*, **30**:517–526.

33. Hao, W., Xianmin, Z., Yongcong, K., Gaofei, O., Hongwei, X. (2013). Solder joint inspection based on neural network combined with genetic algorithm. *Optik*, **124**:4110–4116.

34. Bay, H., Ess, A., Tuytelaars, T., Gool, L. V. (2008). Speeded up robust features. ETH Zurich, Katholieke Universiteit Leuven.

35. Altman, N. S. (1992). An introduction to kernel and nearest-neighbor nonparametric regression. *Am. Stat.*, **46**(3):175–185.

36. Strutz, T. (2016). *Data Fitting and Uncertainty*, 2nd ed. Springer Vieweg.

Chapter 2

Opportunities and Challenges of the Fourth Industrial Revolution

Poonam Jindal[a] and Rakesh K. Sindhu[b]

[a] *Chitkara University Institute of Engineering and Technology, Chitkara University, Punjab 140401, India*
[b] *Chitkara College of Pharmacy, Chitkara University, Punjab 140401, India*
poonam.jindal@chitkara.edu.in

The fourth industrial revolution is a result of increased technology developments in artificial intelligence (AI), machine learning, and robotics paired with high computing power and big data. There has been an exponential rise in this technology because of excessive use of automation in the current digital era. AI is already around us in the form of self-driven cars, virtual assistants, and precise diagnosis of critical diseases and is making way for transformative changes in the domestic and business sectors, especially in the design, production, and distribution system. But while there are many potential profits, the fourth industrial revolution can also bring a number of problems for which nobody is prepared. The upcoming intelligent factories may soon replace humans with intelligent robots that will do all the tasks, contributing to unemployment. The influence of this technology can be seen in many products and services that are fast

Artificial Intelligence and the Fourth Industrial Revolution
Edited by Utpal Chakraborty, Amit Banerjee, Jayanta Kumar Saha, Niloy Sarkar, and Chinmay Chakraborty
Copyright © 2022 Jenny Stanford Publishing Pte. Ltd.
ISBN 978-981-4800-79-2 (Paperback), 978-1-003-15974-2 (eBook)
www.jennystanford.com

becoming crucial to the modern age. The fourth industrial revolution will come with new ideas and significant challenges that cannot be easily set aside. It may help solve some of society's problems, for example, in the form of robots driving vehicles and taking care of young and old people, and may also connect machines with mobile and Internet services. It will also bring a fundamental change in the lives of humans by enabling extraordinary advances in technology and will merge the physical, digital, and biological worlds with big promises as well as potential risks.

2.1 Introduction

As the fourth industrial revolution (4IR or Industry 4.0) is all set to reshape the society, its foundation has already been laid by the three large industrial revolutions that have taken place (Fig. 2.1). The first happened when the steam engine was invented and manual processes were replaced by steam engine–powered automobiles, locomotives, industrial processes, etc. It was a huge step for humans to do things faster. The second industrial revolution came when electricity was invented. Power factories, streetlights, and transportation of electricity using cables, etc., made life easier. The third industrial revolution brought automation within factory floors, helping manufacture goods at a faster rate (Xu et al., 2018).

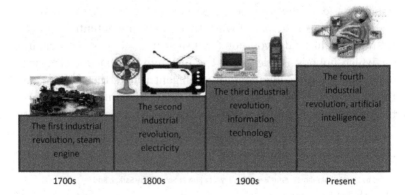

Figure 2.1 Evolution of the first, second, and third industrial revolutions, with their corresponding inventions, followed by the 4IR.

At this moment of time, we are at the cusp of the 4IR, which is going to propel the world much farther because of Internet technology. This is the time when man and machine will come together and build products at a faster rate than anybody has ever seen.

The development of the 4IR is exponential and is affecting almost every country's industrial sector. The entire system of production, distribution, and management would be completely transformed by the scope and wisdom of 4IR. It describes the hiding of the borders between the physical, digital, and biological worlds. It is a blend of advanced artificial intelligence (AI), robotics, the Internet of Things (IoT), 3D printing, quantum computing, and other related technologies (Herweijer et al., 2008). The influence of this technology can be seen in many goods and services that are fast becoming crucial to the present age. We are at the onset of the 4IR, which is exceptional in terms of speed of operation, scale, complexity, and power requirements, unlike the earlier revolutions.

Mechanical production started with item automation in the first revolution. This was followed by mass assembly in the second revolution. The third revolution involved the automation of production, which has highly improved the living standards of people across the world. Undoubtedly, the upcoming sophisticated technology, along with the 4IR, is capable of bringing even more improvements in our lives compared to first initial three revolutions collectively.

2.2 Evolving Fields in the Fourth Industrial Revolution

Some of the upcoming fields in the 4IR are:

- IoT
- Robotics
- Machine learning
- AI
- Nanotechnology
- Quantum computing
- Biotechnology

2.3 Artificial Intelligence: Technology Driving Change

Computers with AI can think like human beings, recognize complex patterns, process information, draw conclusions, and make recommendations.

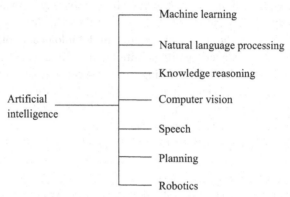

It's a technology that will change the way organizations interrelate and will provide intelligent products and services with new scope (Oztemel & Gursev, 2018). AI is an important part of this overall change and is expected to significantly affect the quality of our daily lives. AI has the potential to create completely new industries that would change the world by providing smart solutions, for example, data analysis would be replaced by unsupervised algorithms.

AI is the advancement of innovation to reproduce and enhance human knowledge. It can be grouped into:

- **Artificial narrow intelligence**: This performs simple single tasks, for example, voice recognition.
- **Artificial general intelligence**: This performs multiple tasks across different areas.
- **Artificial super intelligence**: This is capable of providing intelligence that is beyond human capability.

Today, some big tech companies, like Alphabet, Baidu, Microsoft, Facebook, and Tesla, have their very own aspirations with respect to how they will use AI to convey the maximum utility for their business (Brunette et al., 2009). The AI focus point of each organization will, in general, be different, which will bring about numerous AI applications across various industries, all of which will eventually greatly affect human lives. Here is an example of what the coming time might be like for you: You wake up in the morning with the daily weather report, news, important reminders, etc., available on your smart mirrors. After you finish breakfast, your completely automatic car halts at your doorstep, with the destination set to your office, so that you can sit comfortably in your seat and get ready for the day's tasks. You check your housekeeping robot doing the household chores from your mobile while seated in your office. On your way back home, you use your remote control to switch on the air conditioner at your home for cooling in advance. This is only a small preview of what life may hold for you with the 4IR. AI breathes life into the creative mind as well as changes the manner in which we make the most of our lives (Jeong & Park, 2018).

AI will not only optimize many of our systems but will also create new jobs. It will take away the physical and mental constraints of humans, just like machines that are 100 or 1000 times stronger than humans. AI will work to create different outcomes as never seen before. This technology is capable of:

- Increasing the productivity of internal applications
- Solving complex problems
- Reducing cost
- Enhancing every business process
- Providing smart operations and smart services

AI has already grown in the form of smart machines, drones, 3D printing, smartphones, supercomputers, and ultrasmall chips. Nanoscale devices are already under development, which will create smart factories. In these smart factories, methods of manufacturing would be controlled virtually (Moor Insights & Strategy, 2018).

Some AI converging factors are:

- Big data
- Enhanced processing power

- Global connectivity
- Upgraded algorithms
- Open-source software and data

2.4 Relationship between Artificial Intelligence, Deep Learning, and Machine Learning

Artificial Intelligence is the extensive canopy which covers Machine Learning and Deep Learning. It is illustrated in Fig. 2.2 that deep learning is the subsection of Machine Learning. So each one of them AI, machine learning and deep learning are only the subsections of one ananother.

2.4.1 Machine Learning

Machine learning (ML) is a field of AI with advanced pattern recognition and computational learning hypothesis, while deep learning (DL) is a field of ML. It learns and makes algorithms that can further learn from and make forecasts for sets of data. For example, we can run a program for a busy traffic junction through ML on the basis of data from earlier traffic patterns, and if it has "learned" successfully it will be able to perform better by predicting forthcoming traffic patterns (Australian Law Reform Commission, 2010).

There are two approaches to ML. The first and most commonly used approach is based on statistical modeling, now termed "classical" ML. It is applied usually for resource allocation, trend discovery, forecasting, face recognition, and pricing. The second

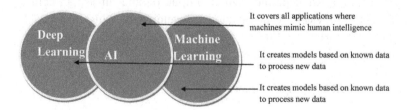

Figure 2.2   Relationship between AI, machine learning, and deep learning.

approach is probabilistic modeling. It sees features and target variables as random variables. The process of modeling shows and alters the level of uncertainty with respect to these variables.

The statistical method is basically suitable for uncovering trends and branches in numerical data and is not dependent on huge data sets (Simon et al., 2016).

ML is applied in the following cases:

- Human proficiency in not available, for example, navigation on Mars.
- Human knowledge is insufficient to extract the complete information, for example, speech recognition.
- The solution varies with time, for example, controlling temperature.
- The result is dependent on specific cases, for example, the biometric system.
- The problem statement is too high for our limited knowledge, for example, Web page rank calculation.

2.4.1.1 Types of machine learning

| Supervised machine learning | Unsupervised machine learning | Reinforcement machine learning | Recommender machine learning |

- **Supervised machine learning**: The program gets training on a predetermined collection of examples that subsequently help it to arrive at a correct result when dealing with a new set of data.
- **Unsupervised machine learning**: The program is provided with a cluster of data and should locate patterns and associations within them.
- **Reinforcement learning**: An agent is trained to perform in an environment and take suitable action so as to be eligible for some reward for taking the correct action and a penalty for a wrong action. It is programmed by many machines and software to take the best possible action in a particular situation.

It is unlike supervised ML in that in supervised ML, the model gets training with the answer key with itself whereas in reinforcement learning, no answer key is provided but the agent has to decide on how to perform in a given task. If the data set is not present, then it has to learn from its own experience (Das et al., 2015).

2.4.1.2 Types of reinforcement learning

Reinforcement learning is of two types:

- o **Positive**: It creates a positive impact on the behavior of the system and increases its strength and performance. It will also maximize the performance and will sustain that for a long period of time. But it has the drawback of overloading the states if too much reinforcement is applied and will spoil the results.
- o **Negative**: It will strengthen the behavior by avoiding or stopping a negative condition.

2.4.1.3 Applications of reinforcement learning

- o In robotics for industrial automation
- o In ML and data processing
- o To build training systems that provide custom instruction and materials according to the requirement of students (Sutton & Barto, 2015).

- • **Recommender learning**: Recommender systems (RSs) are learning techniques that help consumers to customize their sites according to their tastes, that is, help them to search for new services, like books, transportation, music, or people, on the basis of the recommended items. For example, an online customer can get a review or rating of the item which he/she is looking for because of the RS. This system also plays an important part in decision-making by helping the user to increase the profit and reduce the risks. Nowadays, RS is being used by several information-based companies, for example, Google, Twitter, LinkedIn, and Netflix (Portugal et al.,

2015). There are mainly two categories of ML algorithms in RS: content-based recommendation and collaborative recommendation. However, both approaches are combined in modern RSs. These help customers to obtain and withdraw data by making novel and smart recommendations. RSs are mostly used by e-commerce sites (Adomavicius & Tuzhilin, 2011).

2.4.2 Deep Learning

DL, a new branch of ML, aims to move ML near to its actual target, AI. To understand data in the form of images, sound, and text, DL makes various levels of representation and abstraction. DL originates from Neocognitron, which is an artificial neuron network (ANN). Just like the network of neurons inside the brain, ANN is made up of several interrelated processing units. The actual idea was to develop a learning method from the model of the human brain. However, this strategy lost favor among ML groups because it requires unfeasible measure of time and immense data for training the system parameters for any good application. DL can train a multilayer ANN with little data, which brings ANN back into the equation. Taking an example of face recognition to compare DL with ML, whereas an ML algorithm remembers the eyes and nose as parts of the face, a DL algorithm will remember additional features, like the distance between both the eyes and the length of the nose.

The following are some applications of DL:

- Optical character identification, for example, image scanning to extract text from it
- Speech recognition, for example, text formation from a sound clip
- AI, for example, robotics
- Automotive-based applications, for example, self-driven cars
- Military and surveillance, for example, drones

2.4.2.1 Role of deep learning in big data

In today's world, a huge amount of labeled and unlabeled data is being used everyday by various sources, for instance, for big tech companies, like Yahoo!, Google and Microsoft, and social media apps, like Facebook, Twitter, and YouTube, millions of consumers have created enormous amounts of data. These data are beyond the range of conventional database systems and limited analyzing methods in terms of computing, processing, and storage capability. Consequently, information is customized to big data, which have high extension and the capability to change various divisions. But numerous difficulties remain in the way of big data analytics. These challenges can be summed up in terms of the three Vs: volume, variety, and velocity (Nagwa et al., 2016; Banerjee et al., 2019).

- **Data volume**: There is no limit on the size (volume) of the data, that is, the size of the data that are being handled can't be constrained. However, the speed of processing is consistent. Examining and controlling huge volumes of data involves new assets that can appear and show the mentioned outcomes.
- **Data variety**: Data are available in progressively different and complex arrangements from a number of sources and perhaps with various appropriations. To join information that is different in source or structure and do it at a sensible expense is a great challenge.
- **Data velocity**: Quick falling off of continuous data is a major test. Move rates can be constrained, yet demands are boundless. The already existing systems are unable to analyze these continuously moving data.

Data learning and big data are the two main branches of data science. DL algorithms take out complicated data patterns by a learning method hierarchy by analyses and learning a huge amount of unsupervised data, that is, big data. Thus it becomes a highly important means for big data analyzers (Nagwa et al., 2016).

2.4.2.2 Deep learning applications for big data analytics

A hierarchical multilevel learning approach is used for significant abstract representation by DL algorithms. This extracted form can be considered a good basis for big data analytics, like tagging and recognition of data, retrieval of information, and real speech processing.

- **Object recognition**: Computer vision programs make constructive decisions for the genuine objects and scenes from images. Object identification, 3D models, medical imaging, and intelligent cars all come under these programs. The main issue of large-range object identification is to achieve capability in both component extraction and classifier preparing without compromising performance. Feature detection by means of deep networks is highly efficient for carrying out object recognition.
- **Speech and audio signal processing and tagging**: Advanced DL methods have learned to carry out predictions in highly nonlinear problem settings. To attain a good-quality performance, a large quantity of data is required for training the neural networks. Now audio data sets are available in the public. Therefore, there is also an increased demand for tagging labels for them. Only those tagging labels that point out the presence or absence of a particular kind of event for a recording are required and not those containing any worldly information, like weak labels (Morfi & Stowell, 2018).
- **Retrieval of information**: Information retrieval is the method to provide multimedia objects to users in order to satisfy their need for data. Recovery of data at a high rate becomes crucial in big data as there is a huge volume of data in terms of text, audio, video, and images. Several domains contain these multimedia files. Deep networks are mostly used for retrieving information from these files to extract quality features for further processing. Hinton et al. explained a generative DL model for fast information retrieval based on the binary codes of documents (Hinton & Salakhutdinov, 2011).

2.5 AI Challenges by Potential Environmental Areas

AI is considered to be the most important technology for the development and benefit of humanity as well as earth. Figure 2.3 highlights the four potential environmental challenges and the action areas to be benefitted by implementing AI.

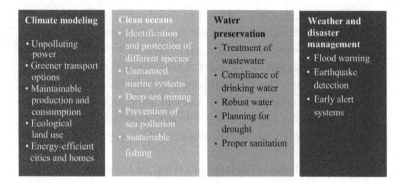

Figure 2.3   Applications of AI to manage the potential challenges in front of society.

2.5.1 Climate Modeling

AI has the capability to deal with climate challenges, for example, by providing the solution for unpolluting power. ML can be applied to meet supply and demand by establishing smart grids to increase efficiency. In Norway, Agder Energy is utilizing AI with cloud to prepare for the varying energy needs, particularly due to the rapid increase in the number of battery-operated vehicles. Neural networks are also used to enhance the efficiency and reliability of renewable power. DNV GL, Norway, provides the data for ML for monitoring capability by using sensors linked to solar and wind-based power stations. This enables the remote inspection of sites and their predictive maintenance, thereby reducing the cost of wind and solar energy (Herweijer et al., 2008).

Data can be analyzed from a number of smart sensors located at remote locations within a building using ML algorithms to estimate the energy requirement and its cost. It can also optimize energy efficiency in buildings by providing lighting and heating from buildings to streets. ML algorithms are widely used to optimize navigation and improve safety and traffic congestion. AI-based automatic vehicles use machine vision algorithms and deep neural networks that will enable a transition over the coming years. Moreover, there will be substantial reduction in greenhouse gases from urban transport with the use of connected autonomous vehicles. Route optimization and services like ride sharing will reduce the driving distance and bottleneck problems.

2.5.2 Clean Oceans

AI, along with other techniques, has helped create new methods to manage and protect oceans by gathering data of locations that are hard to reach as well as ensure sustainability of fish and protection of other species and habitats and observe the effects of changes in climate. To monitor illegal fishing activities, ML algorithms are being employed for accurate patrolling schedule and efforts are on to implement vessel algorithmic patterns for satellite data along with automatic identification systems for ships. Such types of advances will enable authorities to prevent overfishing and control fisheries. Similarly, for the protection of species, image analytics and ML are used to track the numbers and locations of species. Ocean conditions, like water pollution levels, temperature changes, and pH value changes because of climate changes, can be detected by using AI-powered robots. Advanced autonomous technologies, like robotics and nanotechnology, with increased resolution are also being explored to investigate ocean conditions, identify different species, and map resource management. ML computer modeling and satellite imagery is being used by NASA to monitor and predict the present and future conditions of phytoplankton on the surface of oceans (Herweijer et al., 2008).

2.5.3 Water Preservation

Water preservation is critical for the coming generations. Scientists and engineers are working to improve the conditions of reservoirs and usage of water for various geographical areas, along with weather forecast, by employing AI for making better policy decisions. AI is combined with industry intelligence to design smart meters that can identify leakage, monitor water flow in real time, and check whether the meter is malfunctioning. Alerts on the quality of water based on real-time data acquired from ML algorithms can be sent on users' mobiles. AI can also check for aging water pipes in buildings to locate leak-prone pipes to repair and can analyze the data from pressure sensors to estimate the working conditions of filtration systems to reduce the wastage of water. AI incorporated with satellites can forecast weather patterns, which will help to analyze the soil and surface water conditions. This will help predict drought circumstances to help affected sectors and people. Researchers are likewise attempting to join ML with physical models to configure water plans and ascertain capital ventures, drought plans, and potential results of water management choices (Herweijer et al., 2008).

2.5.4 Weather and Disaster Management

The upcoming applications of weather and disaster management are based on advanced information about extreme weather conditions and natural disasters. Predictive analyses governed by AI accompanied by IoT, drones, and sophisticated sensors will help governments and research centers to monitor floods, tremors, and windstorms as well changes in the sea levels and the possibility of other natural risks in real time for automatic triggers that will help timely evacuation if required. AI is used with image processing to process data to give real-time extreme-weather forecast based on social media information. Many meteorological agencies and technical companies work on big data analytics and AI along with more sophisticated physics-based modeling to identify the effect of crucial weather conditions on infrastructure and systems to implement disaster handling policies. Besides this, real-time response planning, DL algorithms, and image processing techniques can be applied for identifying the aging of buildings, materials used, etc. Satellite images can be used to prioritize relief efforts in disaster affected areas (Herweijer et al., 2008).

2.6 Emerging Technologies

4IR, or Industry 4.0, is not just about smart machines and their interconnection. The opportunity it provides is much wider ranging, from big data analysis to augmented reality and from nanotechnology to 3D printing. It is a fusion of all these technologies,

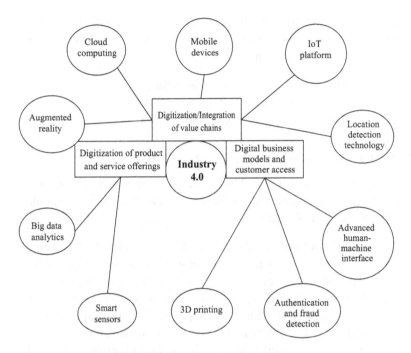

Figure 2.4   Key drivers of the fourth industrial revolution.

as shown in Fig. 2.4, and is spreading much faster and more widely than its predecessors.

2.6.1 Key Drivers

A list of rankings of various technologies behind the 4IR is provided by many organizations. The selection of technologies listed in this chapter is based on the research done by the World Economic Forum and the forum's Global Agenda Councils. All the upcoming technologies and developments have a common feature in that they are under the influence of the universal power of digitization and information technology.

For example, gene sequencing was not possible without the advancements in computing power and data analytics. Similarly, advanced robotics would not happen without AI, which in turn depends on computing power. To identify the drivers of the 4IR, the list is organized into three clusters, all three strongly interconnected by the various technological benefits they offer each other on the basis of the findings and improvements each makes.

- Digitization/Integration of value chains
- Digitization of product and service offerings
- Digital business models and customer access

2.6.1.1 Digitization/integration of value chains

There are four main physical indicators of the technological areas of digitization or integration of value chains: cloud computing, mobile devices, and IoT platform.

- **Cloud computing**: Cloud computing has become the prime enabling factor of the 4IR by opening new opportunities for the business sector to innovate around emerging technologies. Disruptive technologies, like robotics and AI, along with the cloud approach, will help develop new applications and services that will flourish in the future and may take them to a completely new level. For example, industrial equipment designer General Electric (GE) is based on the cloud platform technology that securely accepts and analyzes the data of industrial devices for automation and optimization of the business processes, resulting in a highly efficient supply sequence and also providing advance maintenance (www.oracle.com/uk/cloud/paas/features/next-industrial). New innovative applications based on an integrated cloud platform are making the way for a truly disruptive technology.
- **Mobile devices**: As we move into the new decade, we will see smaller screen sizes combined with computerization and intelligence amplification that will enable us to do a ton of work that we physically do today with a simple tap on a cell phone or a voice direction. In the 4IR, business will be driven by versatile, enhanced, and intelligent programming instead of hardware. In the upcoming age, laborers won't sit behind a work area but will be attached to their mobiles. Thus, there is a need to design simple yet smart programs for applications. One such program is Corpa, a mobile

application to manufacture, incorporate, robotize, and run the key components of business and work processes, reports, dashboards, and undertakings at one spot regardless of the size of the industry. It is a smart mobile app that enables us to run several elements of business without opening any other app.

- **IoT platform**: The key connection between the physical and advanced applications upheld by the 4IR is the IoT, which can be defined as the relationship between things and people, possible due to the connection between technology and various other platforms. Sensors and different methods for interfacing real things to virtual systems are sprouting at a remarkable pace. Compact, inexpensive, and smarter sensors are being installed at homes and in accessories and transport and energy networks, along with manufacturing practices. This will modify the manner in which a supply chain is managed, by monitoring and optimizing the activities at a granular level, and will affect every industry, from assembling to foundation to human services.

2.6.1.2 Digitization of product and service offerings

There are three main physical indicators of the technological areas of digitization of product and service offerings: augmented reality, big data, and smart sensors.

- **Augmented reality**: Augmented reality (AR) is considered the prime factor for the evolving computing power and is the most intuitive and collaborative computing experience. There is a strong connection between IoT and AR, and this is used to extensively analyze data from multiple sensors. Many companies, like GE, have utilized AR extensively in their factories' workflows and processes. For example, in the case of heavy machinery, the internal components that are in need of replacement can be easily viewed. Big companies, like Apple and Google, are developing headsets and platforms based on AR that will undoubtedly shape the next computing platform. The 3D displays shown using AR have certainly amplified the efficiency of people and their ability to focus on work. With Industry 4.0 on its way, it's time to think of how to safeguard our workforce and manage AR to increase the productivity of workers and enhance the prevailing processes and workflows.

- **Big data**: Since we are living in an era where IoT and related technologies are growing quickly, it demands a wide range of data to be processed everywhere throughout the world. Moreover, computer-based intelligence, big data examination, 3D printers, robots, and all such contemporary high-innovation items depend on data collection. Their dynamic application in a wide range of business and life purposes would signify that "information" rules the world, and this is the thing that we call the data-based industrial revolution. Gartner Inc., an American IT research and consultation company, has predicted that big data, AI, and IoT will rise at 15% per annum, which is very high compared to the growth rates of most developed countries. Among these technologies, the use of big data is likely to influence a range of business sectors. For example, insurance companies can launch new services and new products on the basis of data on natural disasters and can set the premium charges on the basis of the geographical data of disaster-hit regions. Similarly, better medical services can be provided on the basis of data collection on medical consultations. Power generation–based companies can also benefit by providing electricity to customers in the most effective manner by accumulating information on power consumption from smart meters.

- **Smart sensors**: With the evolution of various technologies, the industry is converting to smart manufacturing by incorporating IoT and progressing toward the 4IR. For any industry to progress today, it is important for it to capture contextual information, giving more predictive diagnostics, and smart sensor technology can bring a change in the working of industrial devices and help the present as well as future generations of smart machines. Industries have already been using the sensor technology for over a decade, but the use of big data and proficiencies in software have led many business sectors to acquire sensor-based technology so that the ordinary industrial machinery can be replaced by intelligent devices that connect to smart networks. For example, IoT-based sensors along with self-diagnostic and digital circuit expertise are dealing with a variety of processes and industrial requirements, like reducing maintenance cost and improving the effectiveness of equipment. All in all, the industry is moving toward a new generation of smart machines that incorporate sensor-based operations.

2.6.1.3 Digital business models and customer access

There are four main physical indicators of the technological areas of digital business models and customer access: location detection technologies, advanced man-machine interface, authentication and fraud detection, and 3D printing.

- **Location detection technologies**: Advanced technologies, like cloud computing, big data, IoT, and AI, are a part and parcel of the 4IR and are helping the world to adapt as well as adopt and apply the changes in the way we think, function, and make decisions. The industry also focuses on the mobility by embedding maps, geographical information system (GIS), and global positioning system to do things in a different way. Location detection services are more popular than ever, and the industry should concentrate on new apps that work flawlessly. Detection and tracking cannot be considered as a single technology but a combination of many technologies that can be fused together to design systems that help track vehicles, mobiles, laptops, personal digital aids, and gaming consoles and thus their users. As the digital era is reaching new heights, the industry also needs to focus on GIS to connect government sectors, private sectors, academia, etc., so they can collaborate and do things in a different way, resulting in a more supportable world (Schutze et al., 2018).
- **Advanced man-machine interface**: Rapid technological development is being accompanied by significant changes in the human-machine interface (HMI). The demands and tasks are changing for people in the industry, with the more

flexible production systems. Strong technological support by established interrelated technologies can realize the full potential of workers, and they can emerge as decision makers and problem solvers. The HMI started its journey with the ordinary push button and has advanced to touch screens, and many variations are still expected in the way humans communicate with machines. IoT and 4IR will play a significant role in this regard as they are the real adopters of HMI transformation. These advanced HMIs should be highly sophisticated for improved productivity and remote activities, particularly when laborers are in dusty, damp, or dim situations. Keeping these things in mind, new HMIs are being employed by the Industry 4.0 and IoT developers, like advanced touch interfaces, voice interfaces, gesture interfaces, and AR/virtual reality (VR) tools (Papcun et al., 2018).

- **Authentication and fraud detection**: As we progress toward the 4IR and more digital transactions are conducted, individual identification is becoming crucial for humans, devices, legal entities, and others. The proof of identity is a prerequisite for a person to access critical applications or participate in economic, social, and political systems. Similarly, the identity of a device is important for conducting transactions as devices are going to operate independently (without humans) in the near future. With the increase in the data size, the risk of data being stolen and misused also increases. It is assumed that in 2020, 1.7 megabytes of data will be created every second for each individual living on earth. This can reach 163 zetta bytes, or 163 trillion gigabytes, by 2025. Thus it becomes increasingly important to protect data and make sure that the transactions take place in a secure and trustworthy network. Powerful antifraud projects, frameworks, and controls have been created to avoid losses. For example, real-time authentication gives end-to-end authentication and prevents frauds against contact centers. It is based on voice detection and verifies the identity of the caller within a few seconds of the call involving regular conversation with the customer. The key benefit is that the fraudster is exposed and blocked in real time (Klosters, 2018).

- **3D printing**: We are already seeing the impact of the 4IR in our lives, jobs, policies, and industry as our physical and digital worlds are merging. Technologies like AI, robotics, and 3D printing are challenging the manufacturing industry with new ideas and capabilities that were earlier beyond imagination. 3D printing specifically is the one to reshape the region. The 4IR is all about reinventing the methods of manufacturing. Three-dimensional printers have already been making prototypes for a few years, but their actual potential is yet to be explored. In the near future, advanced records will be delivered to any edge of the world, enabling specialists and originators to work from any place without the restriction of conventional assembling methods. Therefore, sooner rather than later, 3D printing will guarantee unprecedented mass personalization and large-scale manufacturing. The digitization of assembling will likewise empower quicker improvement cycles. In light of the fact that there is endless transportation of items and materials forward and backward around the planet, 3D printing will lead to nearby sourcing and assembling, having huge ramifications for exchange, guidelines, and tax assessment (Oztemel & Gursev, 2018).

2.7 The Role of Robotics in the 4IR

Robotics is an area of engineering that describes the design details and applications of robots along with computers in industries to enhance the speed of the manufacturing process. Robots are the basic necessity of any production process in any industry today. Robotics will reduce the need for manpower in the production line with the use of control system and information technology.

Robotics and AI form a powerful combination for automating factory tasks. Recently, AI has emerged as an essential component for finding solutions in robotics, providing flexible learning capabilities in the already existing robust applications. Though AI is still in its blossoming stages, it can be a transformative technology for applications in the manufacturing sector, although there are many that have yet to feel the impact.

2.7.1 Applications of AI and Robotics

AI is applied in the manufacturing industry in the following ways:

- **Assembly line**: AI is very useful in robotic-based assembly lines in industries. It can train the robot to make decisions on its own to decide the best possible path for specific processes while it is in operation. It is also highly beneficial in complex manufacturing processes, like aerospace, by providing real-time correction.
- **Packaging units**: AI-based robotic packaging units are faster, more cost efficient, and accurate. It avoids any breakage of the parts and at the same time improves the overall process by allowing easy handling and movement of robots.
- **Commercial usage**: Robots are now being utilized in retail outlets and hotels across the world. Most of them are based on AI processing in order to make them more like humans to interact with customers.
- **Customized robotics**: In the present era, where almost everything is automated, robots with AI capability are being sold as open-source products. Customers can train their robots to do particular tasks. The bonding of customized robotics with AI could have huge possibilities in the upcoming AI robotics. Smarter, more accurate, and profitable robots can be designed in collaboration with AI. AI has yet to explore its full potential because as AI grows, so will robotics.

Robotics plays an important role in the transformation of various industries. It has simplified traditional processes by replacing humans. It presents a creative method to computerize repetitive tasks without changing existing frameworks and foundation (DevicePlus editorial team, 2017).

There are basically three technical categories to meet the challenges for developing this technology of industrial robots to service robotics or shape the future capability of robots. The first area is knowledge, with which the robot is able to understand, perceive, and plan to enter the actual world. This feature will provide flexibility to the robot to perform independently in different complex conditions. The second area is the ability to manipulate, which provides the robot accurate control and proficiency to manipulate things in its region. This enables it to perform a variety of tasks in various manufacturing processes. The third category is interaction. This is the most important one because it enables the robot to learn to collaborate with humans. It also helps in improving the verbal and nonverbal robot-human communication. The robot has the unique feature of observing and copying the human behavior and learns from experience. Many digital devices have been developed on the basis of this technology, such as computers, mobile phones, tablets, cameras, and radio-frequency identification readers, with improved quality, safety, production, and maintenance, increasing efficiency and effectiveness in all areas (Karabegovic & Husak, 2018).

Moreover, it may be possible to robotize and improve work processes, back-office procedures, and information technology support in the following areas:

- Accounting
- Customer service
- Finance
- Human resources
- Supply chain management
- Telecommunications

Robotics is an area of AI that can enhance the proficiency of various companies. Big companies, like Amazon and Alibaba, are already using robots to perform simple tasks that were otherwise undertaken by humans, which has transformed their business. Even nonretail organizations are working on different humanoid automated robots that can be actualized in a number of situations. The importance of incorporating robotics in various fields lies in the fact that the humans have financial, physical, mental and social needs alongwith the restrictions imposed by certain work laws which puts a restriction in productivity. whereas robots don't have such needs which results in an exponential development in productivity. Thus in coming years, an appreciable number of manual jobs will be replaced by robots (Balalle, R., Ballalle, H., 2008).

As per data analysts, in the coming 30 years, the worldwide economy will change, endangered by imbalance created by AI. As intelligent machines will take over everything, from thinking to flipping burgers, expenses will be cut but social imbalance will increase. Giving robots rather than people a chance to do physical work, the improvement in AI and mechanical autonomy implies that computers will most likely think to an ever-increasing extent. This is seen in the US market that the large number of employment is based on low pay which is an alarming situation.

Despite the downsides, AI and robotics will lead the 4IR. The primary element of this pattern is the combination of technologies. The union of mobile Internet, cloud computing, big data, new energy, robotics, and AI will progressively wipe out the limits of the physical, computerized, and organic worlds. These innovations will meet and open up an entirely different era.

With these advancements in AI and robotics also arises the risk of significant job losses. As per a recent research, around 47% of the jobs in the United States would be lost because of AI, ML, and cognitive computing. A live example of this is the Henn na Hotel, Japan, in which the hotel staff has been replaced by robots that assist customers in checking in. Similarly, the restaurant staff has been replaced in the dining rooms by groups of vending machines. Japan is leading in the field of autonomy and robotization, with around 250,000 modern dynamic robots in service across businesses. Though this does not mean that there will be no jobs for humans, we need to concentrate on creative and high-rated work. It's likewise important for governments around the globe to put extra effort into bringing an essential change in the system of education to best set up the new workers, which will help governments to effectively deal with the disruptions in the worker market in the long haul.

2.8 Conclusion

According to Schwab, the founder and executive chairman of the World Economic Forum, the 4IR is the most vital of the industrial revolutions in terms of speed, scale, technology, complexity, and transformative power. In this chapter, we have discussed the various technological challenges as well as the opportunities that will arise on account of this revolution. Disruptive technologies, like AI, IoT, robotics, AR, VR, ML, and DL are changing the way people live and work and are connecting the physical, digital, and biological worlds to achieve higher gains in industries.

The 4IR should be directed and intended to accomplish expansive additions. Proper planning and guidelines are important to mask the benefits of fourth IR and limit the potential harms that it might cause. Government must take steps to ensure that the countries have the fundamental framework needed to roll out the changes. Most importantly, governments need to take measures to ensure that the 4IR occurs in a manner advantageous for the entire society.

2.9 Future Scope

The 4IR is changing the way we live and work as well as strongly influencing our surroundings. The advancements in AI, robotics, autonomy, and other developing technologies will reclassify the real and virtual worlds we are living in. In the future, there is a need to set up limits that keep the 4IR on a path to profit the entire mankind. One must identify and handle the likely negative effects of 4IR for data safety, privacy, employment, uniformity. and reliability and inculcate positive qualities with the innovations we make, consider how they are to be utilized, and structure them in view of moral application and on the side of collective methods for protecting what's essential to us. It is the moral responsibility of governments, international establishments, organizations, policy makers, worldwide associations, scholarly community, and common society to come together to direct the development of ground-breaking technologies in a manner that restricts the risks and creates a safe, progressive world with common objectives for the collective good of humans in the coming future. The world we make through technological advancements can shape our lives and is the one we will give to the coming generations.

References

Adomavicius, G., Tuzhilin, A. (2011). Context-aware recommender systems. In *Recommender Systems Handbook*, Springer US, pp. 217–253.

Amit, B., Chinmay, C., Anand, K., Debabrata, B. (2019). Emerging trends in IoT and big data analytics for biomedical and health care technologies. In *Handbook of Data Science Approaches for Biomedical Engineering*, Elsevier, Chapter 5, pp. 126–154.

Balalle, R., Balalle, H. (2018). Fourth industrial revolution and future of workforce. *Int. J. Adv. Res. Ideas Innovations Technol.*, **4**(5):151–153.

Brunette, E. S., Flemmer, R. C., Flemmer, C. L. (2009). A review of artificial intelligence. In *Proceedings of the 4th International Conference on Autonomous Robots and Agents*, Wellington, New Zealand, pp. 385–392.

Das, S., Dey, A., Pal, A., Roy, N. (2015). Applications of artificial intelligence in machine learning: review and prospect. *Int. J. Comput. Appl. Technol.*, **115**(9):31–41.

Elaraby, N. M., Elmogy, M., Barakat, S. (2016). Deep learning: effective tool for big data analytics. *Int. J. Comput. Sci. Eng.*, **5**(05):254–262.

Freund, K. (2018). The artificial intelligence starter guide for IT leaders. *Moor Insights & Strategy*, p. 10.

Herweijer, C., Waughray, D. (2018). *Fourth Industrial Revolution for the Earth: Harnessing Artificial Intelligence for the Earth*, PWC.

Hinton, G., Salakhutdinov, R. (2011). Discovering binary codes for documents by learning deep generative models. *Top. Cognit. Sci.*, 1–18.

https://www.alrc.gov.au/publications/9.%20Overview%3A%20 Impact% 20of%20Developing%20Technology%20on%20Privacy/ location-detection-technologies.

https://www.deviceplus.com/connect/robotics-3d-printing/.

Jeong, Y. S., Park, J. H. (2018). Artificial intelligence for the fourth industrial revolution. *J. Inf. Process. Syst.*, **14**(6):1301–1306.

Karabegovic, I., Husak, E. (2018). The fourth industrial revolution and the role of industrial robots: a with focus on China. *J. Eng. Archit.*, **6**(1):67–75.

Klosters, D. (2018). Digital identity on the threshold of a digital identity revolution. *World Economic Forum*, p. 18.

Morfi, D. V., Stowell, D. (2018). Deep learning for audio event detection and tagging on low-resource datasets. *J. Appl. Sci.*, **8**:1397.

Oztemel, E., Gursev, S. (2018). Literature review of Industry 4.0 and related technologies. *J. Intell. Manuf.*, doi:10.1007/s10845-018-1433-8.

Oztemel, E., Gursev, S. (2018). Literature review of Industry 4.0 and related technologies. *J. Intell. Manuf.*, **31**:127–182.

Papcun, P., Kajati, E., Koziorek, J. (2018). Human machine interface in concept of Industry 4.0. *2018 World Symposium on Digital Intelligence for Systems and Machines (DISA)*, Kosice, pp. 289–296.

Portugal, I., Alencar, P., Cowan, D. (2015). The use of machine learning algorithms in recommender systems: a systematic review. *Expert Syst. Appl.*, **97**:205–227.

Schutze, A., Helwig, N., Schneider, T. (2018). Sensors 4.0 – smart sensors and measurement technology enable Industry 4.0. *J. Sens. Sens. Syst.*, **7**:359–371.

Simon, A., Deo, M. S., Venkatesan, S., Ramesh Babu, D. R. (2016). An overview of machine learning and its applications. *Int. J. Electr. Sci. Eng.*, **1**:22–24.

Sutton, R. S., Barto, A. G. (2015). *Reinforcement Learning: An Introduction*. A Bradford Book, The MIT Press, Cambridge.

Xu, M., David, J. M., Kim, S. H. (2018). The fourth industrial revolution: opportunities and challenges. *Int. J. Financial Res.*, **9**(2):90.

Chapter 3

Role of AI in the Advancement of Drug Discovery and Development

Shantanu K. Yadav,[a] Poonam Jindal,[b] and Rakesh K. Sindhu[a]

[a] *Chitkara College of Pharmacy, Chitkara University, Punjab, India*
[b] *Chitkara University Institute of Engineering and Technology,
Chitkara University, Punjab, India*
rakesh.sindhu@chitkara.edu.in

Artificial intelligence (AI) has been successfully used in areas such as computers, voice recognition, and natural language processing and is now swiftly finding its way into areas requiring significant domain expertise, such as biology, physics, and chemistry, the goal being to get better success rates and lower the cost of drug discovery and drug development. The pharmaceutical industry is also using AI to centralize the data sources that were segregated earlier, hire data scientists, and invest in infrastructure. AI's applications in drug discovery are categorized into novel discovery and target identification, hypothesis generation, virtual screening, compound generation, chemical property prediction, ADME/toxicology, prediction of the results of clinical trials, and real-world-evidence-based analysis and actuarial pharmacology. This topic will highlight the latest advances in and perspectives on all kinds of AI technologies used in drug design and development.

Artificial Intelligence and the Fourth Industrial Revolution
Edited by Utpal Chakraborty, Amit Banerjee, Jayanta Kumar Saha, Niloy Sarkar, and Chinmay Chakraborty
Copyright © 2022 Jenny Stanford Publishing Pte. Ltd.
ISBN 978-981-4800-79-2 (Paperback), 978-1-003-15974-2 (eBook)
www.jennystanford.com

3.1 Introduction

Historically, experiments were recorded in paper lab notebooks and then archived for patent purposes. In the preceding 10 years, paper lab notebooks have been substituted by electronic lab notebooks to enhance privacy and for basic information registration, entry, and recovery. Many procedures have been altered in the pharma field for safety purposes, some previous (e.g., Birch reduction—an organic reaction where the aromatic rings are reduced by 1,4 to provide unconjugated cyclohexadiene in the presence of sodium or lithium metal in liquid ammonia and in the presence of alcohol) and some current (e.g., Stille is a chemical reaction commonly used in organic synthesis. The reaction involves the combination of two organic groups, one of which is held as an organotin compound). The collection of portfolios has been introduced into palladium-catalyzed and boron composition (Buchwald–Hartwig and Suzuki responses, among others) and is now commonly used. Transition metal catalysis includes C–H activation responses, which are particularly helpful for rapidly diversifying molecules in the late stage.

We need to turn pharmaceutical development for the best result and a bright future. There has been little change in the manner in which we create tiny molecules in the laboratory. We are still conducting reactions in round-bottom flasks (or crystal pipes), stirring in alternatives (preferably at room temperature), and adding reagents with falling funnels, syringes, or cannulas. The products of such procedures are cooled using microwave generators, and it is still common to heat using hotplates and cool in cold baths. Work-ups too are much the same—quenching the reaction, partitioning in a separating funnel between water and an organic solvent, washing the organic stage, and isolating the products. The general tendency is toward increased use of disposable machinery as contaminated glassware is difficult to clean. Robots in liquid dispensing are used in chemistry laboratories to ensure negligible contamination and mistakes. Different scientists (Campbell et al., 2018) have sought to increase the production of pharmaceutical drugs using AI.

In this chapter, we will reflect on drug synthesis and all pharmaceutical aspects (unearthing and development) of drugs. In the future, this approach will predict changes, target identification, virtual screening (VS) and clinical trials in pharmaceutical science to be developed with the help of various AI techniques.

3.2 Artificial Intelligence

AI is a representation of the mechanism of human intelligence displayed by processors (computers) and has been described to have had a truly revolutionary effect on the production of drugs.

As per the latest reports, machine learning (ML) and big data could have a significant impact on the healthcare system and could ultimately result in a market that generates US$100 billion in annual revenue. Industry professionals believe that medicines can be produced using AI. AI must display animal intelligence, needing to perform duties such as voice recognition, visual perception, language-to-language conversion, and decision making. AI is applicable in various sectors, like government, retail and customer, marketing and advertising, finance, education, transportation, agriculture, aviation, computer science, heavy industry, job research, media and commerce, military, music, publication and writing, online communication and telephone, power electronics, toys and games, and healthcare.

There are so various definitions equilibrium around one's computer code or hardware with human-capable behaviors, stared intelligent, if humans showed it.

Preferably, the term "artificial intelligence" is used to indicate a machine's capacity to imitate human behavioral functions. The primary benefits of AI are the creation of new data, a greater degree of accuracy, autonomous simulations and projections, constant results, and early detection or tracking of a multitude of illnesses. AI engines are generally used to analyze, interpret, or manage information or complicated tasks more accurately. In this sense, AI integrates statistical model identification techniques, biological approaches (e.g., neural networks [NNs]), computational intelligence, and likelihood theories. Pharmaceutical and biotech industries use ML in a number of different ways: for biomarker identification and exploration; for pharmacovigilance (PV) operations, involving adverse event (AE) case analysis or compliance intelligence collection; and for real-world data, which may include the use of large data sets of claims or digital medical data (Russel et al., 2018).

3.3 Machine Learning and Deep Learning in Artificial Intelligence

AI includes ML as a subfield. ML uses arithmetical approaches effective in learning even without explicit programming (Lee et al., 2017). ML is defined as controlled, unmonitored, or improved training. ML, which is defined by mathematical processes that enhance learning with experience, represents the digital portion. ML algorithms are of three types: (i) supervised, (ii) unsupervised (capability to find designs), and (iii) reinforced learning (various software and machines are used to find the best possible behavior or path to take in a specific situation). Overseen/supervised learning includes regression and classification methods in which the analytical system is founded on input and output data. Production from controlled ML includes disease detection under the subgroup classification; medication effectiveness and absorption, distribution, metabolism, excretion (ADME) estimation under the subgroup regression. The outputs from this form of ML include de novo drug design under decision making and experimental designs under execution—where both can be accomplished through modeling and quantum chemistry (Chen et al., 2018). Deep learning (DL) uses artificial neural networks (ANNs) that adapt to a broad range of research data and learn from them. When used separately and in combination, they help in the field of regenerative medicine founded on hereditary markers. An increasing amount of data as well as a steady rise in computer power have led to the advent of DL. The remarkable factor that makes DL an AI subfield is the versatility in the design of NNs, which can be divided into coevolutionary NNs, fully connected feed-forward networks, and recurrent NNs (Grys et al., 2017).

3.4 Application of Machine Learning in Pharmaceutical Science

3.4.1 Disease Identification and Diagnosis

Identification and treatment of diseases is the main goal of ML studies in medicine. According to a 2015 report issued by Pharmaceutical Research and Manufacturers of America, more than 800 drugs as well as vaccines used to treat cancer have been reviewed. In an interview to Bloomberg Technology, Jeff Tyner, a researcher of Knight Institute, said that while this is fascinating, it also creates a challenge to discover ways to work with all the resulting data. "This is where the idea of a biologist collaborating with data scientists and computer scientists is so essential," Tyner said.

3.4.2 Drug Discovery and Manufacturing

The use of ML in the preliminary stages of drug discovery enables, for example, the initial screening of drug substances and the identification of the expected rate of success taking biological factors used in research and development (R&D) tools, such as next-generation protein sequencing. Precision medicine, which includes the identification processes for "multifactorial" disorders or, in effect, alternative routes for therapy, appears to be the frontier in this field. Most of this research involves unmonitored learning, which is still largely confined to the detection of patterns in data with no predictions (the latter is still in the area of supervised learning).

3.4.3 Smart Electronic Health Records

Text classification using both support vector devices and optical character detection (transforming cursive or other handwriting into digitized characters) are the main ML-based technologies to help advance the processing and digitization of digital medical information. Examples of such technologies are MATLAB's ML multitouch technologies and Google's Cloud Vision API for visual object detection (Young et al., 2017).

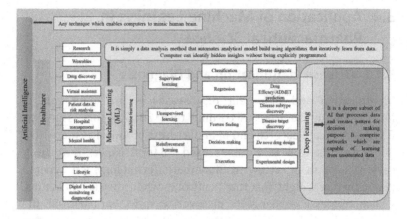

Figure 3.1 Applications of AI and its subpart: deep learning and machine learning, in clinic.

The integration of AI into the healthcare sector is shown concisely in Fig. 3.1. AI is the use of technique that enables a computer to mimic the behavior of humans (Fig. 3.1). Its subfield ML uses statistical methods that have the ability to learn with or without explicit programming. ML is divided into supervised learning, unsupervised learning, and reinforcement (Lee et al., 2017).

3.5 Building an AIF

Compared to many current technologies, the artificial intelligence framework (AIF) is simple. However, it is essential to note that synthetic intelligence is regarded by the Magrathea team as a cobweb rather than a metal chain, implying that the inclusion of individuals, structures, and equipment leads to a substantial possibility of mistakes and losses. Therefore, a lot of effort and dedication is required in a medical environment to understand the data stream within a complex system. At the heart of a slightly complex system is the incorporation of smart learning or "epochs" that are actually difficult to modify and are open to many forms of failure (Yampolskiy, 2016).

Although AI failure may make for fascinating reading or an excellent film, its vulnerability to a strike makes the whole process highly uncertain when you consider that AI and ML techniques are used within the clinic for customers, where correct results and safety are essential. For any successful AIF, the development of era training models within an AI structure is essential.

The Magrathea team used complex systems and DL related to patient medical care, where patients saw the advantage of early disease detection. In a clinical environment, inside clinical genomics, AI plays a vital role in galvanizing tools. To date, little work addresses the "linking" of clinical decision-making moves. The team has a state-of-the-art clinical practice and a traditional hospital environment at Magrathea, operated by the Genatak groups in Kuwait and Nottingham, England. A fresh scheme was required to deal with the enormous amounts of information and clinically appropriate expectations at the start of the personal genome initiative (International Human Genome Sequencing Consortium, 2004) and further animal genome mapping of the next generation for clinical comprehension. The original core initiative was a tiny one, involving the whole genome sequencing (WHGS) of 200 human genomes, which subsequently extended to the WHGS of another 750; in the autumn of 2010, clinical assessment and understanding of the entire information was incredibly hard.

3.6 Classification of Artificial Intelligence

AI is classified into two types, Type 1 and Type 2 (Fig. 3.2).

3.6.1 Type 1

Weak AI or narrow AI: It focuses on one small assignment at a time, and devices that are not too smart can be constructed in a manner so as to appear clever. An example would be a poker match where all laws and motions are fed into the system that fights humans. Every feasible situation must be identified manually before such a match is created. Each and every type of fragile AI helps build stronger AI.

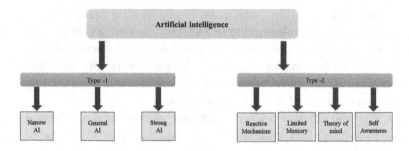

Figure 3.2 Classification of AI.

General AI: It is AI that understands what it is teaching and may use its understanding to attain predefined objectives in fresh and innovative ways. Also, the implementation of this teaching and understanding is not restricted by a framework.

Strong AI: It refers to systems that can execute duties like human beings. There are no adequate current instances of such AI, but many sectors are eager to build such powerful AI, which has led the way to its quick advancement.

3.6.2 Type 2

Reactive machine: It is a basic form of AI. It has no memory and for upcoming actions, it cannot customize past data. An example is the IBM sports program in the 1990s that beat Gary Kasparov.

Limited memory: To guide potential choices, an AI system can use previous events. A few such choice-making structures have been incorporated in self-driving cars. Limited memory is used to analyze and to advise on the next course of action, such as a self-driving vehicle deciding to change lanes. These findings are not recorded incessantly, but Apple collects Chatbot Siri data. A Chatbot is a computer program that allows humans to interact with technology using a variety of input methods such as voice, text, gesture and touch.

Theory of mind: This type of AI can understand and communicate. It can understand people's feelings, opinions, ideas, and aspirations. Not many advances have been made in this sector.

Self-awareness: This AI has consciousness, is incredibly smart, has self-awareness, and is responsive. In short, it is like a complete human being. Of course, there is no such bot as of now and if it is created, it'll be one incredible piece of machinery.

3.7 General Aspects of AI

AI instruments allow you to forecast in vitro reactions; identify new treatment pharmacokinetics, along with their quantitative structure-property relationship and quantitative structure-activity relationship (QSAR); calculate the correct dosage; measure the porosity of hair or the blood-brain barrier, etc. Based on the significance of the pharmacokinetic properties of various medications, the use of silico instruments to predict their properties should lead to enhanced efficiency and decreased cost of drug research. Also helpful are machine training methods, such as support vector machines (SVMs), the Gaussian method, random forest, classification and comparison tree, and Naïve Bayes (NB) classifier. ANNs are capable of creating differential load-output data types, selecting ideal chromatographic gradient environments, analyzing multiple linear regression connections in pharmaceutical testing, predicting drug conduct, and designing preformulations (Sun et al., 2003). NN modeling is a successful method to describe the molecular structure of an organic compound and predict its physical and chemical behavior (Dowell et al., 1999). ANNs are made of computing elements called artificial nerves or nodes. Depending on their functions, ANN designs are categorized into associated, function-extracting, and subadaptive networks (Xu et al., 2011). Because the connection between design and method variables and discharge characteristics of managed-release medication distribution devices is not straightforward, associated networks are chosen to create and optimize managed-demand formulas. In addition to application in genomics, metabolomics, information processing, pharmacy product growth, and assessment of medicinal bioavailability behavior, ANNs were proposed when credible instruments were required to capture source-and-effect interactions and to estimate discrepancies between in vitro and in vivo information (IVIVC). The option of building complicated pharmacokinetic parameter interactions is provided by IVIVC in conjunction with ANNs. ANNs may not be helpful in defining the correlated systems between variables. As a nonlinear modeling tool, IVIVC to design the tablet formulations of the sustained-release paracetamol produced information by a distributed cellular network regression (generalized regression neural network [GRNN]) assessment that was discovered to be similar to what was observed in vitro. This shows the appropriate consistency of a GRNN assessment as well as its capacity to generalize a complicated interaction for both entry and yield dimensions, compensating for variations in medication discharge kinetics under distinct circumstances (Parojcic et al., 2004).

3.7.1 AI Use in Drug Development: R&D Proficiency

Well-established drug-likeness criteria are specifically applied in the production of medicines, but medicinal corporations face significant difficulties in enhancing R&D performance. The price of drug identification/creation has now grown to US$3 billion, from US$800 million in 2001. The cost of producing drugs includes the cost of failures; therefore, the assessed cost is the average value of the novel medicine to be launched into healthcare (Masi et al., 2015). The decreasing rate of the number of new drug molecules authorized per billion US dollars paid on R&D is disturbing. Scannell et al. (2012) methodically examined the main causes of the troubling figures and identified the following main causes of R&D ineffectiveness: (i) the "better than the Beatles" problem, (ii) the cautious-regulator problem, (iii) the elementary research–brute force bias, and (iv) the propensity to throw money into it. According to the better-than-the-Beatles metaphor, a novel molecule for a specific pharmacological disorder should be more effective than any of the currently approved drugs. The challenges discussed here contribute to the fate of any experimental medication promising a better therapeutic result. The stakes are raised with every new drug and the R&D inefficiency is taking place here as it becomes increasingly difficult to clear the ever-progressing hurdles. The cautious-regulator problem refers to the increase in rigor or tightening of controls due to drug calamities of the past. It is an incremental attempt to reduce the acceptance of risk of new medicines and to develop a roadmap for safer product formulations suitable for customers, but this significantly raises R&D spending. This is accompanied by the propensity to throw money into research, a tendency in which R&D institutions hire an increasing number of humans and other resources in the belief that this will ensure a high rate of return and that these institutions will be the first to promote a novel drug. This also results in increasing R&D investment. Clinical research–brute force bias refers to the tendency to overestimate the prospects of basic research developments in the field of preclinical study, in particular target identification or verification and brute-force testing to enhance clinical drug development success rates (Scannell et al., 2012). Rising R&D costs and high turnover levels in the process of creating new medicinal molecules present problems for medicinal companies. Previously, several causes and potential interventions for reducing attrition rates have been evaluated. (The attrition level was represented using the first in-human trials to enroll 10 developed pharmaceutical corporations in the period 1991–2000, in which the daily rate of products making through regulatory production as well as approval was approximately 11% (i.e., 1 in 9 molecules). During Phase IIb and Phase III clinical trials, about 62% of the new chemical entities did not hit the hospital. Nonetheless, these causes are not common, as evidenced by medical literature, which indicates that different treatment areas provide varying success rates. The degree of depletion of molecules with new mechanisms of action is lower than that of compounds with previous mechanisms of action. Pharmaceutical establishments must assess the various clinical areas with differing success rates to evaluate the success of the reduction of attrition rates. Through knowledge of these relative factors, establishments could reduce the levels of attrition (Mullard, 2016).

3.7.2 Application of AI in Drug Designing

The 3D structure of most proteins is calculated by the one-dimensional (1D) sequence of their amino acids; however, while most proteins already have 1D sequence details to be collected, the exact determination of their 3D configurations has not yet been achieved. Many disorders are strongly associated with protein activity disorders. It is possible to use protein-based drug design approaches to create active molecules that target amino acids. Determining the 3D molecular structure is time consuming and costly in laboratories so computer-assisted analysis of 3D protein structures is required.

Prediction of the protein structure is usually broken down into smaller components, such as secondary protein structure, Skeleton torsion angle, solvent surface accessible and so on, which are collection of one-dimensional, structural features (Heffernan et al., 2015). The number of challenges is because of the space of conformation. For large data sets of protein sequences that are available, AI technology has been widely used to predict the structural properties of proteins. Qian and Sejnowski (1988) first used the small cell network system to estimate the secondary composition of spherical acids. Spencer et al. (2015) used the deep belief network in 2015 to estimate the secondary protein composition to a precision level of 80.7%. Wang et al. (2016) eventually incorporated the deep neural system with such a conditional random field as well as suggested a deep convolutional neural field technique for predicting the secondary protein structure. This enhanced the predictive precision to 84%, which could be used to estimate the functional features of proteins, such as touch amount, impaired region, and solubility. In encouraging the growth of this sector, the DL technique has shown excellent potential. DL methods have been used in the preceding years to estimate secondary protein constructions, structural torsion points, α-hydrogen dihedral points, and water-accessible ground regions. For instance, Qi et al. (2012) used deep neural networks (DNNs) as a classifier in 2012 to create a local multitask local protein structure predictor. The protein cellular amino acid sequence is mainly calculated by the two corners of torsion connected to each C-α. Li et al. constructed four DNNs—deep-restricted Boltzmann models with DL architectures, deep-restricted Boltzmann machines, deep recursive restricted Boltzmann machines, and deep-recurring computer networks—to estimate the protein torsion angle. The anticipated remaining touch amount and the torsion angle value allocation obtained from the DNA parts were discovered to be helpful characteristics to improve forecast precision. So this technique proved outstanding in terms of efficiency in the Vital Observation of Protein Structure Prediction (CASP)12 contest.

DeepMind used Alpha Fold to enter the CASP13 contest in December 2018, anticipating a DNA composition with a precision level up to 58%. Once more, such instances indicate DL's excellent capacity for implementation in the forecast of the molecular structure of protein.

3.7.2.1 Protein-protein interaction modeling

In many biological procedures, protein-protein interactions (PPIs) are not crucial but they are linked to several diseases (Scott et al., 2016). The PPI architecture often contains concentrations of protein-protein binding locations, which represents a fresh goal (Santos et al., 2017) that really varies from conventional objectives (ion channels [G-protein], kinases, nuclear receptors) to extend the destination space as well as encourage the growth of tiny molecular medicines. Thus, in addition to the introduction of protein annotation, an in-depth learning of the interface zone of PPIs is important for drug design centered on complicated protein-protein construction and the treatment of associated ailments (Wilson et al., 2013). Many PPI prediction calculation methods have been generated. Due to drawbacks of modern PPI approaches, such as high price, long-term commitments, large quantities of data noise, and strong false positive and negative levels, PPI knowledge is very limited and restricted. The current PPI prediction, therefore, covers mostly two structure- and sequence-based categories. In particular, the method based on the protein prototype structure is more reliable and simpler as most PPI interfaces are conservative. For example, Maheshwari and Brylinski (2016) developed a template structure–based eFindsite PPI prediction approach to identify residues of PPIs from a weak homologous template. This approach is predictively accurate both in the experimental protein configuration and in the protein, structure created by silicon. The PPI interface can be modeled by using the protein-protein docking model on the basis of the complementary theory when the 3D configurations of the two interacting proteins are identified (Vakser, 2014). Structure-based prediction algorithms conduct better than sequence-based techniques, even with limited quantities and performance for recognized protein structures. For example, presently known bacteria, yeast, or humans contain little organizational data for 80 percent of PPIs. AI has made important strides in anticipating PPIs using sequence-based techniques, with exponential growth in protein sequence information (Mosca et al., 2013). Du et al. (2016) used integrated hidden Markov models (ipHMMs) in 2016 to obtain Fisher fractional characteristics from protein sequences. A stacked autoencoder was used to build a DNN perfect for estimating protein residues in an interaction matrix. The DL model's general forecast precision is 80.82%, that is 15% greater than that of the old-style ML model.

Recently, to forecast the putative protein complex, Zeng et al. (2018) created an advanced contact Web server centered on the replication technique. It first studies the protein structure homology and constructs two clusters of multiple sequence alignment; it then utilizes the coevolutionary assessment and ResNet process to estimate interprotein interaction. The above technique decreases the need for homology of the protein sequence as well as dramatically increases the precision of the forecast.

3.7.2.2 Virtual screening

VS is among the primary computing drug discovery techniques for detecting effective tiny protein molecules that adhere to drug receptors. It can be used in drugs still under development to filter out additives that comprise improper skeletons and even as an effective technique for fresh hits. Therefore, VS has become a significant supporter of high-throughput screening (HTS), which has low-success and high-cost problems (Lavecchia et al., 2013). Ligand-based virtual screening (LBVS) does not quite depend on 3D protein functional data; its techniques are mainly used to estimate effective ions when there is no destination design and the spatial precision is small. Modern ML techniques, like k-nearest neighbor (k-NN), boosting, SVM, RF (data fusion), decision trees, DL, and NB, were commonly used in LBVS, not just for efficiently enhancing the frequency of anticipated hits but also for reducing the frequency of fake hits. A strong instrument of further enhancing LBVS is the development of DL, with automatic extraction and layer-by-layer teaching characteristics. Xiao et al. (2018) built an open-source tensor flow DNN model of big data and used it as an LBVS instrument for screening large databases of molecule. By screening 95 million medicines in PubChem before 2015, its system recognized 50% of the drugs after 2015 with a fake positive rate of just 0.01%–0.09%, which proved the enormous capacity for the implementation of DNNs in LBVS.

Structure-based virtual screening (SBVS) is usually used after tests or computational modeling has explained a target's 3D structure. This technique is primarily used to investigate the relationships between feasible effective ligands and proteins of binding sites and generally demonstrates stronger predictive efficiency than the LBVS methods (Arciniega et al., 2014). However, because of the amount of protein structures and highly complex protein conformations, the SBVS-based technique faces the issue of exponential growth. Pereira et al. (2016) used profound teaching to enhance SBVS efficiency and used the deep convolutional neural network (DCNN) template to construct an enhanced SBVS technology called deepVS. This technique requires the outcome of molecular docking as the DCNN entry and therefore can automatically derive and discover from the fundamental information appropriate characteristics such as molecule form, nuclear relative charge, and nuclear range.

Skalic et al. (2019) also recently introduced the DCNN model to construct a Web service called BindScope that can identify dormant and effective large-scale compounds and boost the graphics processing unit (GPU). DL's outstanding efficiency in separating nonbinding and binding protein-ligand interactions is exactly what has significantly encouraged the fast growth of VS.

3.7.2.3 Quantitative structure-activity relationship

Mathematical techniques are being used to build quantitative modeling connections between the physicochemical or chemical characteristics of drugs or even between their biochemical activity.

The QSAR technique primarily involves information compilation and pretreatment, molecular descriptor formation as well as choice, mathematical design institution, analysis, design assessment, and model application (Myint & Xie, 2010). Highly potent lead compounds can be found during lead optimization by analyzing or anticipating the action of a sequence of drugs. When AI can build a robust model of the link between biological action and chemical structure effectively, it becomes an essential fragment of QSAR.

As early as 1990, Aoyama et al. applied neural networks (NNs) to QSAR analysis. Subsequently, various traditional ML methods, such as RF, Boosting, GP, *k*-NN, DL, Cubist and SVM have also been widely used to construct QSAR models. The QSAR system is becoming increasingly complicated, and it is hard to satisfy the requirements of collecting large amounts of information with the superficial NN technique used during traditional ML.

Dahl et al. (2014) established a multitasking DNN in 2014 that can immediately forecast a compound's chemical and biological properties from its chemical structure. In addition, Winkler and Le (2017) discovered that DNNs should produce a stronger QSAR model in the event of sparse data collection as they can convert the entry descriptor into some kind of higher-dimensional space, thereby efficiently enhancing their evaluation efficiency and the continuous cliff issue experienced by QSAR in Fig. 3.3.

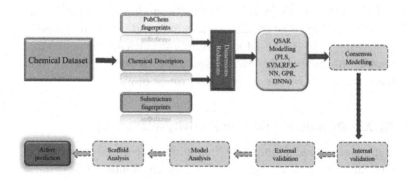

Figure 3.3 Workflow of QSAR modeling. Copyright 2016 Springer.

3.7.2.4 Assessment of ADME

Given the discovery of millions of active compounds over the past few years, the number of potential molecular units permitted by the Food and Drug Administration (FDA) has not increased over the years (Xue et al., 2018). The primary reason is that such active compounds do not follow medication requirements in terms of ADME. As per statistics, the key causes of drug development loss were weak pharmacokinetics (39%) and preclinical toxicity (11%), in addition to the lack of effectiveness, commercial reasons, adverse effects, etc. Improvement in the rate of success in drug development and manufacturing output is largely dependent on effective calculation and stabilization of the ADME of leading complexes. Toxicological studies in vivo are now the benchmark for evaluating drug-induced side effects. Therefore, ADME based on computers has become a method of choice for advanced drug detection. Different models have shown success in the estimation of ADME properties, and it's even suggested that intensive use of computational technology could reduce drug development costs by up to 50%. The accuracy of the computer-prophesied ADME has increased primarily due to the availability of high-quality data and robust numerical analytical techniques. The system structure has progressively changed to a nonlinear multivariate approach based on AI algorithms, from the original linear multivariate models, such as partial least squares and multiple linear returns (Tan et al., 2010). While epoxidized metabolites always lead to drug toxicity, accurate estimation of the site of epoxide formation should effectively decrease the risk of metabolite creation and help acquire safer medications. Hughes et al. (2015) used DCNNs and 702 databases of epoxidation reaction to create a model to predict the site of epoxide (SOE) accurately. The design integrates various epoxidation responses empirically, which accomplishes SOE predictions at the atomic scale, with an area under curve (AUC) of up to 95% and an AUC of 79%, respectively, for epoxidated and nonepoxidated molecules. Moreover, they merged the undirected graph recurrent neural network molecular coding construction with the equivalency to construct a predictive model called DL drug-induced liver injury (DILI) based on 475 drugs. When this model was applied to the external data set of 198 drugs, the AUC was as high as 0.955, which was substantially higher than the former DILI models (Xu et al., 2015).

3.7.2.5 Drug repurposing/drug reposing

Drug reposing is described as a way to find new signs from drug molecules that have been licensed. Reuse of drugs for novel functions will not only decrease the price of developing drugs but also efficiently lower the hazard of drug safety problems. Identification of drug-target interaction (DTI) is a vital part of drug designing and repurposing (Chen et al., 2016). Nevertheless, there is little data available for drug repurposing research, and the requirement of large-scale DTIs cannot be satisfied by specific bio-analysis approaches. Predicting DTIs with computational methods has become a significant research direction. The two most widely used prediction approaches in DTI are ligand-based approaches and structure-based approaches, as shown in Fig. 3.4. Constructed on the postulation that molecules with the same structure can have different physiological activities, the ligand-based methods apply the QSAR to model the target molecules' biological activity (Duran et al., 2014). Methods based on structure primarily use molecular docking to monitor molecules according to the target's crystal structure. These two conventional analytical approaches have been significantly hindered by the small number of recognized mark active molecules as well as the 3D structure of mark proteins. Several ML methods have been applied in the AI processing of

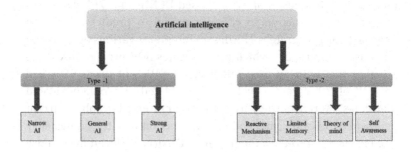

Figure 3.4   Flowchart of deepDTIs (Banerjee et al., 2019).

heterogeneous data to predict DTIs. A classification model with a drug-target pair as the input and whether there is an interaction between drug-target pairs as the output are the currently used ML methods. The most common ML models are bilateral structures, such as RF, SVM, and ANNs (Duran et al., 2014). DL methods have attracted much greater attention due to their good performance and the ability to learn abstract data representation at multilevel. Mayr et al. (2005) assessed the quality of several DL approaches on a huge set of data on R&D of drugs, contrasting their findings with the findings of other conventional ML approaches for predicting drug targets. They found that feed-forward DNNs significantly outstrip all competing approaches and their performance is comparable to or even better than that of in vitro assays in predicting specific targets.

Incorporating heterogeneous data fonts, such as the relation between drug and disease, may help to improve the reliability of DTI forecasting. To predict new DTIs through heterogeneous networks, Luo et al. (2017) developed a computational process known as DTINet based on this relationship. In addition to incorporating heterogeneous data (such as proteins, drugs, illnesses, and opposing effects), this also effectively tackles the noisy, incomplete, but high-dimensional features of a large-scale biological source by studying the characteristics of drugs and proteins to reflect small-dimensional but information-rich vectors. Experiments have also confirmed the novel relationship between the three drugs prophesied by DTINet as well as the cyclo-oxygenase enzyme, displaying a comparatively durable prognostic presentation of DTIs.

3.7.2.6 De novo drug design

De novo drug development is based on an algorithm that requires a molecular strategy and an analysis machine to create innovative chemical entities. In general, compounds formulated using this approach have low pharmacokinetics properties and are difficult to synthesize. By comparison, a wide digital library of chemical structures is created using the ligand-based process. A score function was used to scan the chemical area, considering drug metabolism and pharmacokinetic (DMPK) properties, feasibility of synthesis, biological activity, and similarity of question structure. Therefore, it is possible to obtain several synthetically possible molecules (Schneider et al., 2018).

Another method is to model an analogous question construction based on the conversion rules of medicinal chemists. Conversion can help generate new compound structures efficiently and reliably. In order to improve the current problems of de novo drug design, the DL with its strong generalisation and learning capacities, the technique has been used to produce automatically new chemical entities with certain predicted properties.

Nonetheless, the latter approaches are production models based on ligand and sequence. Freshly generated compounds have relatively little variation in structure and therefore can usually be obtained by basic chemical adjustments. To this end, in 2018, the growth of the implementation of a deep procreative model in de novo molecular generation was deliberated and exactly summarized and some major challenges of implementing the model in this specific field were highlighted.

- **Use of AI in pharmacology**: The "one disease–multiple targets" model is based on the "one disease–one target" model because of a detailed consideration of physiological pathways in molecular diseases. One disease–multiple targets is referred to as polytarget pharmacology (Zhang et al., 2013). Other databases, such as PubChem, ZINC, Ligand Expo, KEGG, Drug Bank, STITCH, Binding DB, ChEMBL, PDB, and Super target, are available to provide knowledge on binding affinities, drug targets, crystal structures, disease significance, biological pathways, chemical properties, and physiological actions. AI could be used to probe these databases to design polypharmacological agents. Recently, the AI system made a great achievement in the development of polyphar-macological agents: researchers developed a computational framework, deepDDI, for greater consideration of drug-drug interactions and related pathways as well as the prediction of alternative drugs without undesirable health effects (Ryu et al., 2018).

- **Use of AI in population selection for clinical trials**: An appropriate AI model in clinical trials would recognize the disease in a person, classify the target genes, and determine the impact of a molecular layout as well as active and inactive target effects. In a Phase II phase study of subjects with mental illnesses, a new AI tool, known as AiCure, was developed as a mobile application, and it was reported that AiCure increased adherence 25% compared with the traditional modified directly observed therapy. A critical process is the choice of patients for a clinical trial. Detaining the relationship between human-relevant biomarkers and in vitro phenotypes allows for a much more reliable, quantifiable evaluation of the volatility of a particular patient's therapy responses. The development of AI methods for the detection and prediction of human-relevant disease biomarkers enables the recruitment of a specific patient group in clinical trials in Phases II and III. Predictive modeling of artificial data in the selection of tests for hospital patients should increase the level of study performance (Deliberato et al. 2017).

- **Use in postmarketing surveillance**: In 1962, some infants were discovered to have been born with malformed legs when thalidomide, a minor sleeping pill, was taken by pregnant women who often took off-label medicine. The World Health Organization (WHO) International Drug Monitoring System was established following the tragedy. Since 1978, Sweden's Uppsala Monitoring Center has been the regional coordinator for PV in cooperation with WHO and has 134 full member countries with national agencies promoting AE reporting structures for patient health and drugs. These programs show that safety measures are minimal in clinical trials and product safety must be monitored carefully over the lifetime of the drug on the market. Naturally, spontaneous reporting systems such as the FDA event reporting system, have allowed data removal approaches to recognize statistical associations between AEs and drugs (Sakaeda et al., 2013). Nevertheless, with known weaknesses, such as conflicting biases and under-reporting, the focus also shifted to alternative information outlets and novel analytical methods that may replace or complement current resources. Next, we address the techniques and data tools that can support postmarketing PV, along with the relevant AI-driven methods required to collect knowledge and learn from it (Ridings, 2013).

- **Use of robots to monitor and treat**: Robots may also help assess improvements in human behavior in such circumstances as rehabilitation (Silver et al., 2016). AI can help track drug delivery to targeted organs, tissues, or tumors. It is exciting, for instance, to hear about the recent phenomenon of nanorobots programmed to solve distribution issues that arise when difficulties are encountered in spreading the healing agent to a site of importance. This problem arises when the doctor tries to reach a tumor's center that appears to be anoxic and less vascular but most proliferatively active. Scientists have tried to harness a biological agent of desirable properties as a replacement for "intelligent" nanoparticles to address limitations in mechanical or radioactive robotics. They are researching a specific type of marine bacteria, *Magnetococcus marinus*, that naturally travels to low-oxygenated areas (anaerobic), for this purpose. An external magnetic origin provides initial instructions, and then endogenous nanorobot characteristics are put into operation. Such nanorobots can be covalently related to beneficial properties of nanoliposomes. Most of these novel applications of AI in medicine require further study, especially in the areas of human-computer interactions. In 1970, Moshimo Mori presented the concept of an uncanny valley where the human-robot interaction area was an important theme. These experiments evaluate humanoid robots for their apparent humanity, airiness, and beauty (Waltz & Buchanan, 2009). They have been assessed as appropriate.

- **Partnership between pharmaceutical and AI companies**: With the quick adoption of AI in hospitals, particularly in 2016 and 2017, multiple pharmaceutical corporations have taken funds and have partnerships with AI companies in the hope of developing better healthcare technologies. These include developing diagnostics and biomarkers, discovering drug targets, and designing new drugs. The shift from general medicine to traditional AI healthcare is centered on data. Subsequently, the analyses of these underlying data in conjunction with ML and DL are framed into algorithms, thus contributing strongly to broadminded modern healthcare that incorporates AI. Therefore, enterprises have frequently been recentered on a global scale between pharmaceutical productions and AI companies. For example, Google's subsidiary DeepMind Technologies has worked with the Royal Free London National Health Service (NHS) Foundation Trust to help manage acute kidney injury (Powles & Hodson, 2017). A national initiative that uses data and AI from NHS patients with rare diseases is the 100,000 Genomes Initiative in the UK, in partnership with Roche, Berg, Merck, and Biogen.

Atomically, healthcare AI is visionary and is the first DL platform for the discovery of new chemical compounds. Renowned for its exceptional speed, precision, and versatility in structural chemicals, Atomwise has worked to identify potential new therapies for 27 disease goals and partners alongside top colleges, including Stanford University, Yale University, and pharmaceutical companies. Benevolent AI is one of the widely used methods for drug development. It uses text mining to analyze accessible patients and knowledge of genetics and biotics so that the interrelationships between them can be extrapolated to create highly educational charts consisting of complicated maps with over a billion connections. The resultant map, composed of complex interactions, is intended to provide facts to identify new relations or data holes that can lead to new theories. AI follows the illness-agonist section, which makes unbiased views of technical data to produce rational hypotheses, as opposed to human interactions, where the element of possible discrimination is present.

A range of such collaborations cover different exploration zones, such as the documentation of fresh trivial molecules, the development of novel therapeutic approaches, and tracking of the health information through wearable machineries. These innovations are anticipated to lead to better medical services, improved quality in clinical studies, and improved healthcare. To bring modern products to the market, about US$3 billion have been invested over the course of 15 years. This pattern is unsustainable, and changes are essential because customers do not want to pay more for medicinal products and losses. A change in the structure of the company is thus necessary, and AI offers this opportunity. The partnerships show the value of AI engineering in helping us find much better room for development and find unusual particles. This is beyond conventional observations, unless we use only standard HTS for our products (Lexchin, 2012).

3.8 Challenges and Limitations of AI

AI has demonstrated success in defining and categorizing features in and removing features from multifaceted, heavy-dimensional, or noisy data and has helped develop various drugs. Nonetheless, AI faces several problems that have not been tackled effectively. AI techniques are often alluded to as "black boxes." The models are low in interpretability and accuracy, typically have few methods to describe the appearances, and are devoid of sensible descriptions for the related biological mechanisms, and this is a barrier to exposing the integrated biological mechanisms in the evidence used for data analysis. However, in the absence of manual feedback, they cannot disclose the complexity of molecules and structural associations typical in biochemistry.

The second drawback is the requirement of big data sets and the overfitting. AI, particularly DL, usually requires large data sets of training. The size and value of data as a data mining application can significantly affect the reliability and act of the linked models. Nevertheless, the greatest amount of biological information obtained by pharmaceutical companies is usually kept in a personal, proprietary estate in the country, and most important data sets are not available upon request. When the data set is not big enough, DL's major challenge is to fix overfitting possibilities, that is, when there is a minor training difficulty and sample errors are huge, it is difficult to summarize the details included in the data set appropriately. The third drawback relates to the selection of the system and the improvement of parameters. There are many AI model architectures, especially DL models, and new constructions are constantly being proposed. So, selecting suitable models according to the research task's requirements is not simple. Although there are currently some auxiliary selection methods, such as software for optimizing hyperparameters, the entire system mechanism is relatively complex. Moreover, the model of NN training involves extensive adjustment of parameters. Relevant useful guidance, however, is limited, and a broad theoretical scheme has not yet been established to optimize these designs.

Lastly, there is the cost estimation issue. Although AI models involve less computing cost in exercise, the training process is typically time consuming and computationally intensive, particularly the DL model, which has many more hidden layers. GPU is also essential for supporting the processing of some large data groups, leading to comparatively high computing costs. Google's TensorFlow (http:/TensorFlow.org) is a new open-source platform that makes DNN design and debugging significantly simpler. Nevertheless, conventional ML models continue to play a crucial role, particularly when the data volume of the research task is fairly small or the quantity of variables is high. SVM or ensemble learning can be a more fitting design option in these cases.

3.9 Conclusion

There are presently some established drugs that have used AI methods, but it is possible that it will take another two to three years for more drugs to be established on the basis of advances in technology (AI). Experts believe strongly that AI will forever change the drug industry, including how medicines are found. Nevertheless, the person must know how to train algorithms, needing domain expertise, to be effective in biopharmaceuticals utilizing AI. An acceptable environment will be when AI and medicinal chemists work closely, as the former can help analyze huge data sets and the latter can train machines. Specifically, the

DL method is able to handle complicated tasks with massive, high-dimensional data sets without any physical intervention, which has proved valuable in works and business applications. The only way to fully incorporate several large platform data repositories would be to combine ML, particularly DL, with experience and human expertise. AI technology's powerful data mining capability has given a new energy to software-aided drug design, accelerating and encouraging the drug development process. It is expected that AI software will infiltrate many fields of drug development, and new-drug research will thus become a machine-aided form of drug discovery. Combined with the sequential control of automation and smart synthesis software, a smart drug development system may emerge that combines analytics—the AI prediction model—and automated synthesis. Besides, the present situation of the long cycle of drug development, high price, and high failure rate is expected to change.

References

Amit, B., Chinmay, C., Anand, K., Debabrata, B. (2019). Emerging trends in IoT and big data analytics for biomedical and health care technologies. In *Handbook of Data Science Approaches for Biomedical Engineering*, Elsevier, Chapter 5, pp. 126–154.

Aoyama, T., Suzuki, Y., Ichikawa, H. (1990). Neural networks applied to pharmaceutical problems. III. Neural networks applied to quantitative structure-activity relationship (QSAR) analysis. *J. Med. Chem.*, **33**(9):2583–2590.

Arciniega, M., Lange, O. F. (2014). Improvement of virtual screening results by docking data feature analysis. *J. Chem. Inf. Model.*, **54**(5):1401–1411.

Campbell, P. J. (2018). U.S. Patent No. 10,083,453. U.S. Patent and Trademark Office, Washington, DC.

Chen, X., Yan, C. C., Zhang, X. T., Zhang, X., Dai, F., Yin, J., Zhang, Y. D. (2016). Drug-target interaction prediction: databases, web servers and computational models. *Briefings Bioinf.*, **17**(4):696–712.

Dahl, G. E., Jaitly, N., Salakhutdinov, R. (2014). Multi-task neural networks for QSAR predictions. arXiv preprint arXiv:1406.1231.

Deliberato, R. O., Celi, L. A., Stone, D. J. (2017). Clinical note creation, binning, and artificial intelligence. *JMIR Med. Inf.*, **5**(3):24.

Dowell, J. A., Hussain, A., Devane, J., Young, D. (1999). Artificial neural networks applied to the in vitro-in vivo correlation of an extended-release formulation: initial trials and experience. *J. Pharm. Sci.*, **88**(1):154–160.

Du, T. C., Liao, L., Wu, C. H., Sun, B. L. (2016). Prediction of residue-residue contact matrix for protein-protein interaction with Fisher score features and deep learning. *Methods*, **1**(110):97–105.

Duran, F. J. R, Alonso, N., Caamano, O., García-Mera, X., Yaneez, M., Prado-Prado, F. J., Gonzalez-Díaz, H. (2014). Prediction of multi-target networks of neuroprotective compounds with entropy indices and synthesis, assay, and theoretical study of new asymmetric, 1,2 rasagiline carbamates. *Int. J. Mol. Sci.*, **15**:17035–17064.

Heffernan, R., Paliwal, K., Lyons, J., Dehzangi, A., Sharma, A., Wang, J., Sattar, A., Yang, Y., Zhou, Y. (2015). Improving prediction of secondary structure, local backbone angles and solvent accessible surface area of proteins by iterative deep learning. *Sci. Rep.*, **5**(1):1–11.

Hughes, T. B., Miller, G. P., Swamidass, S. J. (2015). Modeling epoxidation of drug-like molecules with a deep machine learning network. *ACS Cent. Sci.*, **1**(4):168–180.

International Human Genome Sequencing Consortium (2004). Finishing the euchromatic sequence of the human genome. *Nature*, **431**(7011): 931.

Kraus, O. Z., Grys, B. T., Ba, J., Chong, Y., Frey, B. J., Boone, C., Andrews, B. J. (2017). Automated analysis of high-content microscopy data with deep learning. *Mol. Syst. Biol.*, **13**(4):1–15.

Lavecchia, A., Di Giovanni, C. (2013). Virtual screening strategies in drug discovery: a critical review. *Curr. Med. Chem.*, **20**(23):2839–2860.

Lee, J. G., Jun, S., Cho, Y. W., Lee, H., Kim, G. B., Seo, J. B., Kim, N. (2017). Deep learning in medical imaging: general overview. *Korean J. Radiol.*, **18**(4):570–584.

Lexchin, J. (2012). Models for financing the regulation of pharmaceutical promotion. *Global Health*, **8**:24.

Luo, Y. L., Zhao, X. B., Zhou, J. T., Yang, J. L., Zhang, Y. Q., Kuang, W. H., Peng, J., Chen, L., Zeng, J. Y. (2017). A network integration approach for drug-target interaction prediction and computational drug repositioning from heterogeneous information. *Nat. Commun.*, **8**(1):573–579.

Maheshwari, S., Brylinski, M. (2016). Templte-based identification of protein–protein interface using eFindsitePPI. *Methods*, **15**(93):64–71.

Masi, A., Quintana, D. S., Glozier, N., Lloyd, A. R., Hickie, I. B., Guastella, A. J. (2015). Cytokine aberrations in autism spectrum disorder: a systematic review and meta-analysis. *Mol. Psychiatry*, **20**(4):440–446.

Mosca, R., Ceol, A., Aloy, P. (2013). Interactome3D: adding structural details to protein networks. *Nat. Methods*, **10**(1):47–53.

Myint, K. Z., Xie, X. Q. (2010). Recent advances in fragment-based QSAR and multi-dimensional QSAR methods. *Int. J. Mol. Sci.*, **11**(10):3846–3866.

Mullard, A. (2016). Parsing clinical success rates. *Nat. Rev. Drug Discov.*, **15**:447.

Parojcic, J., Svetlana, I., Zorica, D., Milica, J., Owen, I. Corrigan (2007). An investigation into the usefulness of generalized regression neural network analysis in the development of level A in vitro–in vivo correlation. *Eur. J. Pharm. Sci.*, **3**(4):264–272.

Pereira, J. C., Caffarena, E. R., dos Santos, C. N. (2016). Boosting docking-based virtual screening with deep learning. *J. Chem. Inf. Model.*, **56**(12):2495–2506.

Powles, J., Hodson, H. (2017). Google DeepMind and healthcare in an age algorithm. *Health Technol.*, **7**(4):351–367.

Qi, Y. J., Oja, M., Weston, J., Noble, W. S. (2012). A unified multitask architecture for predicting local protein properties. *PLoS One*, **7**(3):1–11.

Qian, N., Sejnowski, T. J. (1998). Predicting the secondary structure of globular proteins using neural network models. *J. Mol. Biol.*, **202**(4):865–884.

Reddy, A. S., Zhang, S. (2013). Polypharmacology: drug discovery for the future. *Expert Rev. Clin. Pharmacol.*, **6**(1):41–47.

Ridings, J. E. (2013). The thalidomide disaster, lessons from the past. *Methods Mol. Biol.*, **947**:575–586.

Russell, S., Dewey, D., Tegmark, M. (2015). Research priorities for robust and benefcial artifcial intelligence. *AI Mag.*, 105–114.

Ryu, J. Y., Kim, H. U., Lee, S. Y. (2018). Deep learning improves prediction of drug–drug and drug–food interactions. *Proc. Natl. Acad. Sci. U.S.A.*, **115**(18):E4304–E4311.

Sakaeda, T., Tamon, A., Kadoyama, K., Okuno, Y. (2013). Data mining of the public version of the FDA adverse eventreporting system. *Int. J. Med. Sci.*, **10**(7):796–803.

Santos, R., Ursu, O., Gaulton, A., Bento, A. P., Donadi, R. S., Bologa, C. G., Karlsson, A., Al-Lazikani, B., Hersey, A., Oprea, T. I., Overington, J. P.

(2017). A comprehensive map of molecular drug targets. *Nat. Rev. Drug Discov.*, **16**(1):19–34.

Scannell, J. W., et al. (2012). Diagnosing the decline in pharmaceutical R&D efficiency. *Nat. Rev. Drug Discov.*, **11**(2):191–200.

Schneider, G., Geppert, T., Hartenfeller, M., Reisen, F., Klenner, A., Reutlinger, M., Hähnke, V., Hiss, J. A., Zettl, H., Keppner, S. Spankuch, B., Schneider, P. (2011). Reaction driven de novo design, synthesis and testing of potential type II kinase inhibitors, *Future Med. Chem.* **3**:415–424.

Scott, D. E., Bayly, A. R., Abell, C., Skidmore, J. (2016). Small molecules, big targets: drug discovery faces the protein–protein interaction challenge. *Nat. Rev. Drug Discov.*, **15**:533–550.

Silver, D., Maddison, C., Huang, A., Guez, A. (2016). Mastering the game of Go with deep neural networks and tree search. *Nature*, **529**(7587):484–489.

Skalic, M., Martínez-Rosell, G., Jimenez, J., Fabritiis, G. (2019). Play-MoleculeBindScope: large scale CNN-based virtual screening on the web. *Bioinformatics*, **35**(7):1237–1238.

Spencer, M., Eickholt, J., Cheng, J. L. (2015). A deep learning network approach to ab initio protein secondary structure prediction. *IEEE/ACM Trans. Comput. Biol.*, **12**(1):103–112.

Sun, Y., Peng, Y., Chen, Y., Shukla, A. J. (2003). Application of artificial neural networks in the design of controlled release drug delivery systems. *Adv. Drug Delivery Rev.*, **55**(9):1201–1215.

Tan, J. J., Cong, X. J., Hu, L. M., Wang, C. X., Jia, L., Liang, X. J. (2010). Therapeutic strategies underpinning the development of novel techniques for the treatment of HIV infection. *Drug Discov. Today*, **15**(5–6):186–197.

Vakser, I. A. (2014). Protein-protein docking: from interaction to interactome. *Biophys. J.*, **107**(8):1785–1793.

Waltz, D., Buchanan, B. G. (2009). Automating science. *Science*, **324**:43–48.

Wang, S., Peng, J., Ma, J. Z., Xu, J. B. (2016). Protein secondary structure prediction using deep convolutional neural fields. *Sci. Rep.*, **10**(6):1–11.

Wilson, A. J., Murphy, N. S., Long, K. (2013). Azzarito, Inhibition of α-helix-mediated protein-protein interactions using designed molecules. *Nat. Chem.*, **5**(3):161–173.

Winkler, D. A., Le, T. C. (2017). Performance of deep and shallow neural networks, the universal approximation theorem, activity cliffs, and QSAR. *Mol. Inf.*, **36**(1–2):1–7.

Xiao, T., Qi, X., Chen, Y. Z., Jiang, Y. (2018). Development of ligand-based big data deep neural network models for virtual screening of large compound libraries. *Mol. Inf.*, **37**(18):1800031–1800037.

Xu, W., Li, N., Gao, C. (2011). Preparation of controlled porosity osmotic pump tablets for salvianolic acid and optimization of the formulation using an artificial neural network method. *Acta Pharmacol. Sin.*, **1**(1):64–70.

Xu, Y., Dai, Z., Chen, F., Gao, S., Pei, J., Lai, L. (2015). Deep learning for drug-induced liver injury. *J. Chem. Inf. Model.*, **55**(10):2085–2093.

Xue, H. Q., Li, J., Xie, H. Z., Wang, Y. D. (2018). Review of drug repositioning approaches and resources. *Int. J. Biol. Sci.*, **14**(10):1232–1244.

Yampolskiy, R. V., Spellchecker, M. S. (2016). *Artificial intelligence safety and cybersecurity: A timeline of AI failures.*, **18**(14):1610.07997.

Young, J. D., Cal, C., Lu, X. (2017). Unsupervised deep learning reveals prognostically relevant subtypes of glioblastoma. *BMC Bioinf.*, **18**(11):381–385.

Zeng, H., Wang, S., Zhou, T. M., Zhao, F. F., Li, X. F., Wu, Q., Xu, J. B. (2018). ComplexContact: a web server for inter-protein contact prediction using deep learning. *Nucleic Acids Res.*, **16**(1):17–32.

SECTION II

INTERNET OF MEDICAL THINGS (IoMT)

Chapter 4

Internet of Health Things: Opportunities and Challenges

Emeka Chukwu, Lalit Garg, and Ryan Zahra

Department of Computer Information System, University of Malta, Msida, Malta
lalit.garg@um.edu.mt

This chapter highlights the information revolution and the related opportunities and challenges. It specifically looks at the inherent problems with healthcare systems that stem from the presence of multiple actors and how digital systems have attempted to solve them. The critical problem of the shortage of skilled personnel at the primary healthcare level has led to undercollection or zero collection of vital statistics and biostatistics of patients. The chapter further discusses the emergence of autonomous systems with control systems and the full spectrum of digitization, with specific focus on the Internet of Health Things (IoHT). The key applications of IoHT are discussed, along with the key opportunities and pitfalls and what to look out for when architecting one. The architecture, network, power, and other design considerations of a maternal health use case model are discussed. The full prototype components of a maternal health self-service kiosk are described to illustrate this.

Artificial Intelligence and the Fourth Industrial Revolution
Edited by Utpal Chakraborty, Amit Banerjee, Jayanta Kumar Saha, Niloy Sarkar, and Chinmay Chakraborty
Copyright © 2022 Jenny Stanford Publishing Pte. Ltd.
ISBN 978-981-4800-79-2 (Paperback), 978-1-003-15974-2 (eBook)
www.jennystanford.com

4.1 Introduction: Health System

Historically, across the globe, health has been about the patient getting well and remaining well. Most of the care provisions happen at the health facility, though trends are changing with the emergence of patient-centered care models [1]. There are over a hundred specialties and subspecialties in medicine and many more allied medical and support specialists, like nurses, midwives, pharmacists, and laboratory scientists and their subspecialists. This means that modern healthcare depends heavily on teamwork and communication and information [2]. Also given that healthcare has to do with human life, it is one of the heavily regulated sectors, just as it is complicated. The complex nature of this endeavor has made efficient, equitable, and effective healthcare delivery a "wicked" problem to solve for many countries. Specifically, many countries in the sub-Saharan African region still struggle with issues like preventable deaths of women in labor or immediately after delivery [3]. Routine health information in these countries is still aggregate-based, and other information is augmented by statistical analysis of extensive surveys.

4.1.1 Health Information Revolution

Traditional data collection, use, storage, and retrieval systems are paper based, though efforts are on to digitize these processes and records as much as possible. One significant limitation of the paper-based system is that only one person in the care continuum, in most cases at a different geographic location, can use the paper form or folder at any given time. It can even result in further delays if the patient has multiple reasons for visiting a care facility, as is often the case for the elderly. This limitation and other limitations of paper-based health and medical information have been identified with advances in technology.

Digitization of health facility information, specifically patient information, has led to an enormous amount of information from many different sources. Information is also provided by nontraditional sources, such as imagery equipment, genomic data, and environmental and body sensors. It is now commonplace to have patients access and manage information related to their own health through mobile applications.

4.1.2 Health Workforce and Task Shifting

The issues around human resources for health and their number, quality, and how to track and manage them have occupied the interests of policy makers across many jurisdictions. Recent findings from a few countries show that Africa still lags in terms of the required number and quality of health workers per citizen. Often, health facilities are staffed by untrained volunteers who perform more functions than they are supposed to because of the paucity of the workforce to do these tasks. This development even prompted some countries to recognize and try to formalize these task shift endeavors. Volunteers were provided support to do the work of nurses where there is a shortage of nurses, and midwives could perform some surgical procedures with little support and training. These efforts received a mixed response in terms of success and acceptance amid the cry for ethical consideration. This chapter advocates shifting tasks that machines can do to machines and that patients can do to them. This, it is anticipated, will free up more skilled professionals to deliver better care and help improve health outcomes where they matter most. Also, the data currently collected by these professionals often lack in quality and usability. In many cases, data are collected only for reporting, and no thought is given to closing the feedback loop in an actionable and patient-specific way; feedback happens at the aggregate level only. Audits are few and far between, but when they do happen, they focus on validating that the information entered into the summary form tallies with that in the patient registers. The registers are many, often one for each health program area. For example, there will be a register for HIV/AIDs, another for antenatal, and yet another for family planning.

4.1.3 Digitization: Hope, Hype, and Harm

Technology advances have given rise to a long list of promises that digitization will deliver, and these have been slow to arrive. Health systems in many parts of the world (especially the third world) remain unreliable, unsafe, unsatisfying, too expensive, and out of reach of many [4]. People have been waiting for when and where these promises will be delivered. Some have even taken the initiative to incentivize mass adoption, like the US government's meaningful use project [5]. Many have developed national strategies, yet this often results in little progress. When this information should be collected, in what format, by whom, and where are some of the questions that currently limit the impact of the expected information revolution. Advances in technology have also been known to cause harm and sometimes death in the case of diagnostic systems. The privacy of patient information is attracting interest and sometimes poses impediments to digitization.

4.2 Internet of Health Things

The technology that the Internet of Health Things (IoHT) is based on is the Internet of Things (IoT), a technology that involves connecting several smart devices [6, 7]. The enabling technologies for IoT are mobile communication and wireless networks, which make up the core foundation. Near-field communication (NFC) and radio-frequency identification (RFID) are two of the fundamental technologies that enable IoT [8]. The NFC technology came to fruition as the RFID technology evolved. Nowadays, it is also becoming a norm to include the NFC technology in smartphones, and this offers a different type of wireless connectivity than that offered by the more common Bluetooth [6]. These technologies make it easier and more efficient for devices to collect more substantial volumes of data than was previously possible from different devices and analyze it [9]. The same concept has also been applied to the area of healthcare in many different ways but always to allow smart objects to communicate and therefore create an ecosystem that can provide better healthcare [10].

Before moving forward in this chapter, it is vital to discuss the history of IoT to clarify IoHT, a new term not as popular as IoT but whose popularity is already growing. One may have heard of terms like Web of Things, WoT in short, and Internet of Medical Things, also known as IoMT. This chapter is focused on IoHT. Borrowing the definition from Guinard and Trifa, the "Internet of Things is a system of physical objects that can be discovered, monitored, controlled, or interacted with by electronic devices that communicate over various networking interfaces and eventually can be connected to the wider Internet" [11]. One will agree that many of things that featured in the definition are already around us; this was not the case two decades ago.

Our everyday lives can be extended through the use of digital augmentations. The actual breakthrough was mentioned in the 1965 paper by Intel's chairman emeritus, Gordon Earle Moore, where he predicted that the computing power (or, more precisely, the number of transistors on a silicon wafer) for a given size would double every two years [12]. This prediction seems to have held for the last five years, resulting in microprocessor-controlled devices unprecedentedly small in size. This miniaturization has led to many advances and has increased the world of what is now possible. Today, we live in a world of wearables and embedded sensing devises. Around us, washing machines, televisions, fridges, air conditioners, and any**thing** one can think of can be digitally controlled. When these things are connected, we have the **Internet** (or Intranet), which completes the IoT paradigm. At this point, it is essential to distinguish between the Internet and an intranet, though they both contribute to complete the IoT paradigm. An intranet is a functional network that does not necessarily connect to the Internet.

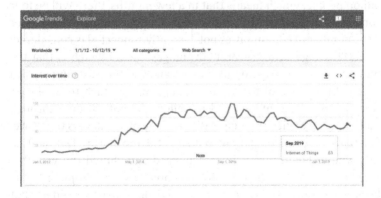

Figure 4.1 Google trend analysis: Internet of Things.

The Internet, on the other hand, is a network that somehow allows access to the Internet worldwide Web.

Describing the different protocols and components that make up the Internet part is well beyond the scope of this one chapter. One clear thing is that our Bluetooth, Wi-Fi, ZigBee, 2G, 3G, 4G, 5G, infrared, and cable connected protocols all make up the network suite. A Google trend analysis shows that though the term was coined sometime in 2009, it did not generate global interest until after 2013, trigged likely by Google's purchase of NEST (an IoT company) for US$3.3 billion in December 2013. The growing interest is shown by the Google trend analysis in Fig. 4.1.

4.2.1 Opportunities

The use cases of IoT in healthcare are growing, with new opportunities for application emerging. While the existence of advanced digital solutions in a health system may make adoption easy, it can often be a barrier to entry. Sometimes, not having existing digital systems, as is the case in most African health systems, can be an opportunity as there is nothing to tear down because nothing currently exists. The opportunities continue to grow with autonomous systems in which machines are now increasingly talking to each other. The IoT market will be worth about US$1.7 trillion in 2020, with Cisco suggesting that the number of connected devices will grow to 50 billion around the same time [13]. This scenario will be in a world with less than 8 billion people, which means that in a few months, there will be four connected devices per individual. This number may even be higher if the number of communities not currently connected is considered.

IoHT brings about several advantages, one of which is that the burden on the nurses to collect and analyze the vast amounts of data is taken care of by this technology. Before such technologies were adopted, the nurses had to collect the patient data through their medical devices manually [8]. In many places across the world, manual collection of patient data is still the norm. The shift to this type of healthcare has also changed and shifted the way that the sector operates. Before such technologies were applied to the healthcare field, the system was based on a reactive model, but thanks to technology, this has shifted to becoming a predictive model

[8]. As observed in Paschou et al.'s study, even professionals in the healthcare industry are utilizing such applications since this may also help them monitor and manage the patient's health remotely. Often, patients themselves are taking up this new technology to improve their own health [10]. The utilization of such wearables and the use of smart devices together with medical devices has also made way for the improvement of the healthcare services and also reduction in the time of hospital stays for patients [8].

4.2.2 Applications

In healthcare, the connected systems vary, like wearables, ambient assisted living, and health facility support instruments. Applications of IoHT are based and focused more on the user, and there are countless examples of applications of IoHT, some of which we will discuss now. Wearables are normally used to monitor physical activity. They are also effective in measuring the temperature and other vital signs using an array of sensors. Wearables can be embedded in, for instance, smart contact eyeglasses for people who are visually impaired [14]; smart wristwatches for determining temperature, oxygen saturation, calculate calorie burn rate, etc. [15]; smart shirts [16]; smart bracelets; smart rings for pulse monitoring [17]; heart rate monitors leveraging accelerometer and sensors [18]; smart belts [19]; and caps [20]. The electroencephalogram cap, for instance, is used to measure electric brain pulses and use the information to understand the triggers for epileptic seizures in patients who use them [20]. For blood pressure measurements, a special kind of sensor-based oximeter is calibrated to read the pulse from a user's finger. These sensor data are then transmitted as blood pressure readings to an on-site or off-site location. The key factors in the effectiveness of the system are the power requirements, consumption, recharge options, and storage, all in the context of size and utility. An IoT-enabled insulin pump is one other innovation that has recorded success in practice [21].

As stated earlier, most of the applications allow for the monitoring of specific vital signs in the human body, most of the time through the use of a wearable device or a medical device connected to a smartphone. Istepanian et al. proposed a system that tracks and senses the patient's glucose levels in real time [22]. Other systems are also available, and most of them make use of a smartphone or a computer connected to a blood glucose collector. Blood pressure is monitored extensively by many people, and by combining a blood pressure reader with a smartphone, one can gain more insight and better analyze the data that have been gathered [23].

Apart from blood pressure, the body temperature is sometimes another vital sign that one would need to monitor [39], and therefore implementations using IoT related to monitoring body temperature have been developed. A body temperature monitor is connected to another system that tracks and analyzes the body temperature.

Another vital sign that can be monitored is the cardiac activity of a patient. Electrocardiogram (ECG) monitoring can be combined with IoT to maximize the benefits, as was observed in several studies [24–27]. This monitoring combined with IoT is based on having an ECG machine connected to a device that processes the data being received, and it flags abnormal activities in real time. One would also want to monitor the blood oxygen levels/saturation, and implementations using IoT that monitor the patient's blood oxygen saturation have been developed, as observed in the study by Jara et al. [43].

Apart from monitoring, some people are required to use and manage some form of equipment, such as wheelchairs, and therefore wheelchair management has also been taken into consideration and systems based on IoT have been developed [40]. In the study by Kolici et al. [44], a system using IoT has been implemented that can detect the current status of a patient using a particular wheelchair while also controlling the vibration in the chair.

4.2.3 Describing a Maternal Health Use Case

In developing countries, where over 90% of the maternal deaths happen, there is opportunity to leverage existing IoT-based systems to collect, manage, store, and aggregate maternal health information for better decision making [28, 29]. The health-worker-to-client ratio remains grossly inadequate, and even available health workers are overburdened with manual collection and reporting of data, which usually end up not being used for decision making. This ratio is wider at the primary healthcare (PHC) level, leading to increased costs in rural areas. Traditional technology solutions have failed to make a dent in developing countries because of the poor infrastructure in terms of water, electricity, and network connectivity, which are often nonexistent. The dream of any data to ease the work of the already overburdened health workers continues to elude practitioners.

The question then will be how IoHT presents an opportunity. I will attempt to provide a theoretical description of the use case here and later give insight into the modeling decisions. This use case will rely on collecting as much information as possible about

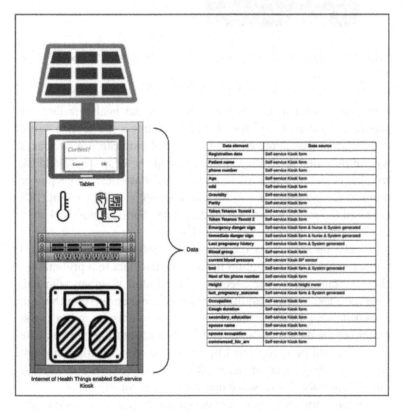

Figure 4.2 Hypothetical IoHT-enabled maternal health kiosk at a PHC.

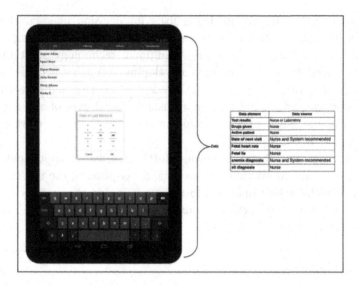

Figure 4.3 Hypothetical reduced data for health workers to capture in the use case.

a pregnant woman long before she visits the health facility for prenatal checkups or delivery. It will work like a self-service kiosk, and sensor-enabled low-energy meters and data-capture interfaces will be provided at strategic locations and the hospital. This kiosk will be powered using solar panels and batteries. Since power remains fundamental, this system will be 100% direct current (DC) powered to reduce losses and inefficiencies. For information that an IoHT-enabled kiosk can collect, see Fig. 4.2.

As can be seen from the figure, the self-service kiosk can aid the patient enter information from biometric data to a preliminary clinical exam before she meets a clinical staff member, who will then have less work in addition to a recommender system through a decision support algorithm. These data elements are only a cross-section of the data that can be collected up-front from patients. The actual list will be determined by the jurisdiction's health ministry. Now, let us look at what the health provider (referred to here as nurse) will track for the same prenatal health domain.

Similarly, the recommender system will help generate additional information, as in Fig. 4.4.

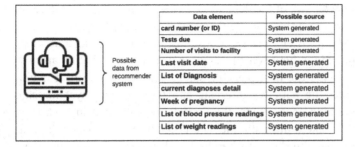

Data element	Possible source
card number (or ID)	System generated
Tests due	System generated
Number of visits to facility	System generated
Last visit date	System generated
List of Diagnosis	System generated
current diagnoses detail	System generated
Week of pregnancy	System generated
List of blood pressure readings	System generated
List of weight readings	System generated

Figure 4.4 Hypothetical data from the recommender system linked to the use case.

4.2.4 Challenges

This rosy use case and its potential health impact will seem self-evident at first, given that the necessary ingredients have been implemented and used in the last two decades. Also, the prices are coming down, so this should not be difficult to implement. But in reality, to deploy this system, many huddles must be overcome, some not related to technology or healthcare but in the realms of behavioral science. How systems are thought of in developing countries needs to be reassessed. Assuming the infrastructure related to electric power is solved, the problem of physical security of the infrastructure still needs to be addressed. One way maybe to design a system that cannot be reused easily without significant technical effort, hoping that this will discourage criminals if they consider it a worthless piece of public good. Deploying a system like this requires a complex interface of multiple engineering, computer science, embedded systems, information technology, game theory, and behavior change. The system will have to be designed to scale to be power efficient.

4.2.5 Limitations

The limitations when implementing IoHT are many, mainly related to privacy, security, and ethical issues, due to the sensitive data that is going to be handled [8]. Ethical issues arise against the risks that come when connecting medical devices to the Internet, as well as the sensitivity of the data, already mentioned with the effect on the healthcare systems on deploying such systems [30].

When developing such systems, one must make sure that the connections between the devices are secure, reliable, and robust to provide the best possible safety and security [30]. It is vital to protect the patient's interests and trust in such systems since if something goes wrong, the confidence and trust in such systems would take a dramatic hit. The other important aspect is to make sure that the data are handled and processed correctly and in a secure way to ensure the interests of those involved [30].

4.3 Modeling an IoHT

This section will describe the service implementation of IoHT and components and modules of a primary healthcare use case.

4.3.1 Service Implementation

IoHT enables a variety of services that, in the past, were not possible without such technology.

One such service is ambient assisted living, which helps to improve the healthcare of the elderly people and those suffering from disabilities. The goals of such systems are to try to maximize the independence of these individuals in their homes, which can be achieved by offering them assistance in the problems they encounter [31]. One such problem that such a patient could face is related to the medication that he or she has to take, and a system has been proposed in Ref. [32] using the RFID technology to eliminate this problem or help with it by controlling the medication using IoT.

Another service made possible by IoHT aims to aid with an adverse drug reaction (ADR), which is an injury caused by taking medication—either the wrong medication or the wrong dose [6]. Systems implemented to prevent such cases make sure that each patient is given the right medication in the right dose by utilizing the NFC technology together with barcodes to identify the drugs related to the patient. Then an algorithm processes whether the same patient has any allergens to the respective drug while also checking the medical records [33]. Yang et al. observed one such system, where a product called iMedBox developed by iMedPack targets an ADR by implementing the RFID technology [24, 34].

A different service also enabled and made possible through IoHT is children health information. This service is aimed at educating and making the society, the children included, aware of the different types of problems that children may face, from psychological problems to mental health problems. A system has been implemented to empower the hospitalized children by educating them and amusing them directly from the hospital wards, as stated by Vicini et al. [35]. A service that aims at educating children about their diets is achieved through m-health (mobile phones) with the aid of their guardians and educators [36].

Wearable Device Access is an application on smartphones that allows patients to monitor their health where sensors in smart devices connected to the smartphone gather the various medical data needed for such applications [37]. Different implementations of such systems that help in monitoring the healthcare of the patient can be observed in Refs. [38, 39].

The architecture of IoHT systems often derives from the open systems interconnection (OSI) architecture, as detailed in Fig. 4.5. One can easily see here that often when IoT or IoHT is discussed, it is in the context of the physical layer of the OSI layers and the Internet protocols suite (Transmission Control Protocol [TCP]/Internet Protocol [IP]) stack.

The physical layer on the OSI stack represents the hardware—the sensors that sense information from the environment and transfer it to the higher layers of the stack through the data and network layer. The power and other physical infrastructures, like the Ethernet and Wi-Fi nodes, are all in the physical layer. All the protocols that can be used to achieve successful transport of the data to the application layer are detailed. Nowadays, the application layer can vary from an operating system (OS) in the embedded device to an OS connected to a phone through Bluetooth, Wi-Fi, or infrared. A successful deployment of the use case will require the detailed implementation of the use case. So far, the kiosk being described

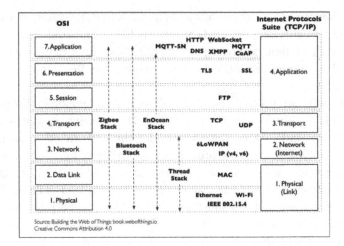

Figure 4.5 OSI and Internet protocols suite (TCP/IP) stack mapping [40].

does not seem any different from conventional kiosks that ATMs and other vending self-services depend on. The novelty of this kiosk will be described now.

4.3.2 Electric Power Module

The entire stack in this conceptual kiosk will be powered by a solar source connected through a solar charge controller, which will then pass direct electricity to the storage battery and the load. To ensure that the kiosk has minimal technical disruption, a notification will be configured to keep track of the solar activity. For instance, a notification can be configured to keep track of the solar panel and battery interface and status. It will ensure that issues are flagged as soon as they occur and are addressed. As the end users will not have the technical capability to interpret the readings, they will be interested in whether the device is working or not. The battery will be modeled taking into account the duration the kiosk is expected to be up if the source is not running. All components in the kiosk will draw DC. To improve efficiency, the central control unit of the application will be located in the raspberry pi and other devices will connect to it through a hub interface. The pi connection and other interfaces are described in detail in Section 4.3.4. It is expected that

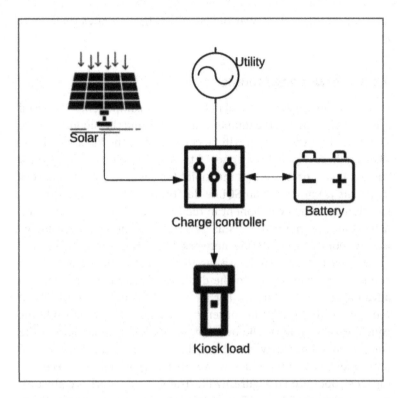

Figure 4.6   Electricity architecture in the proposed kiosk.

on a sunny day, when ultraviolet light is available, it will ignite the panel, charge the battery, and power the load. When the weather is cloudy, the load will draw power from the battery as needed. Each of the components will be calibrated to draw not more than 2 amps at 5 volts. For efficiency, the devices will be configured to use only the energy they require. In locations where there is grid or occasional electric power generator usage, the charge controller will be a hybrid that can take solar and utility supply. It will, by default, be configured to prioritize solar energy if both exist. Figure 4.6 describes this in detail.

This will ensure that the battery remains charged with any available source of electricity, primary or secondary. The battery to be used will depend on the overall power requirements and the duration of tolerable power outage that a backup power supply is needed.

4.3.3 Networking Module

In the conceptual kiosk model in Fig. 4.2, the components discussed are the weight meter, the temperature meter, and the blood pressure meter. Each of the meters will be connected to the kiosk database. The database could be in one of two forms, either hosted on a custom board or hosted on the tablet for client entry. The android stack has a library for an SQLite database [41] and supports this functionality. Each component in the IoHT kiosk will be connected using the multiple protocols as described in the OSI stack under the network blocks. At the network layer, Bluetooth, Wi-Fi, and networked Ethernet cable are the three most probable options to connect the components to the database. Each network layer has advantages and limitations. For instance, Bluetooth and Wi-Fi will require an additional transmitter for each component and will thus consume more power. The best option will be to cable-connect the components since they are all on the same physical kiosk. It will limit energy consumption and overheads associated with keeping the kiosk operational. Figure 4.7 illustrates the network architecture with connectors like RJ45 network interface cables using the IP stack.

The transport mechanism will be TCP, which complements the IP, which is the reason the suite is often called TCP/IP. TCP is routinely used to establish and maintain a network connection, which application programs can use to exchange data. The competing alternatives will be Wi-Fi and Bluetooth.

4.3.4 Server Architecture

When looking at an IoHT device or a suite of devices such as the one we have been discussing, it is essential to look at some characteristics that will influence how these devices operate— features such as the make of the central processing unit; the clock speed (measured in MHz or GHz); the flash memory, often referred to as the random-access memory; read-only memory storage; and

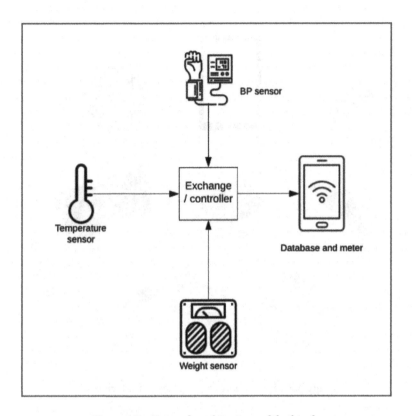

Figure 4.7 Network architecture of the kiosk.

the radio connection (e.g., IEEE 802.15, Wi-Fi, and Bluetooth). Embedded devices are small and very resource-constrained (or limited, whichever you prefer). That is the reason the OSs for these embedded devices are optimized, resulting in OSs like TinyOS [42]. An interface for sensor-enabled systems that runs the system can be developed using a raspberry pi development board [43]. Figure 4.8 shows a sample array of components that can be connected to achieve this server architecture.

We have to choose an embedded system to prototype our server that hosts the database. For this prototype to be helpful to the reader, the raspberry pi prototype board has been chosen as the embedded system to host the server. The server is modeled to be a node.js server (details in Section 4.3.5). The raspberry pi can then also

Figure 4.8 Server hardware and sensor interfaces.

be configured as a Wi-Fi access point that can be connected to a nearby android or IOS-based tablets running a preconfigured mobile application with mobile forms.

Four weight sensors, each calibrated for 50 kg, will be used together to model a 200 kg sensor interface for a 200 kg weighing scale. The oximeter will measure an individual's pulse when clipped to that individual's finger. The pulses will then be calibrated through an analog-to-digital converter to obtain the blood pressure readings. The temperature sensor will read the temperature of the area of measurement. The keyboard will allow users to enter form information, like biodata, into the kiosk. The sensors will read I formation for authenticated users. The boost converter is the interface between the solar charge controller and the raspberry pi. The boost converter will help to convert the solar charge controller electricity output to the appropriate level for the server. This is

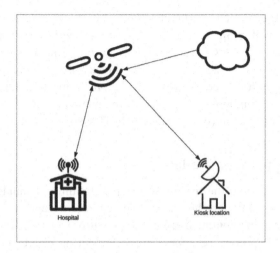

Figure 4.9 Terrestrial satellite connection option.

necessary as the voltage from the charge controller fluctuates and stabilizes at 12 volts. Alternatively, the charge controller can be configured to deliver the voltage needed by the various components of the kiosk. If one is interested in the details of how this is achieved, an excellent article by Fathah will help [44]. Similarly, collected data can be backed up and uploaded to the cloud through a global system for mobile communication (GSM) module. The cloud in the context refers to a central location where data are stored. It could be an on-premise server at, say, a central government location or a remote server hosted by a cloud service provider. The details of this construct will not be the focus of this chapter. The GSM module could easily be replaced by a low-cost broadband satellite solution, like the one depicted in Fig. 4.9. This alternative will have its pros and cons; one advantage is that it can lower the cost of monthly access. But a disadvantage may be increased power consumption. However, the bandwidth-to-cost ratio may justify the alternative. The cheapest satellite Internet provision in terms of cost and form factor is typically the Ka band. The Ka band satellite is notorious for its weather sensitivity and thus cannot be used for mission-critical deployments. However, it offers high bandwidth throughput.

To further the commitment to low energy consumption for this kiosk, we will adopt an E-ink paper display with a gray scale for the display with a lower refresh rate than traditional monitors [45]. This completes the description of the kiosk hardware components. This description excludes discussions on casing and finishes as this description looks at a prototype using available electronic components to achieve an invaluable IoHT system.

4.3.5 Application Module

The next step will be to choose the programming language from the many available in the market. The emergence and widespread use of the REpresentative State Transfer Protocol and Application Programming Interface is increasingly endearing Web technologies to developers . The language of choice has to support Web standards and protocols fully. Such a language should also have a broad user base and support. It will translate to availability of developed libraries for many Hypertext Transfer Protocol tasks, as this can speed up development and reduce errors. It also helps to try and use one language for the entire solution, ranging from the kiosk database interface to the cloud backup engine. JavaScript is the leader when it comes to Web technologies and meets all these criteria. However, JavaScript has been used for client scripting, while other programming languages have been used for the server side. Developers have long wished to work with and learn only one language, especially with the popularity of the Web. The research in this area led to the formation of the node.js foundation in 2015, with IBM, Microsoft, and Intel as leading supporters [46]. Node introduced and democratized the use of JavaScript for server-side programming. Unlike the traditional approach, where an application is deployed in a running server, in node.js, the application is the server.

As stated earlier, the architecture of the application layer of our kiosk for IoHT system will be based on node.js. inspired by Guinard and Trifa [11]. The node.js framework will enable us to build the servers running in the cloud or on the devices and access the physical peripherals connected to our kiosk server. Now that we have decided on the programing language of the application module, let us design the application architecture on which our node.js code will be based. The user walk-through and actions will be described in detail for this design.

4.3.6 User Journey

The first step in designing the node application that supports and manages the server and the IoHT kiosk is detailing the users and use cases. The primary user of this kiosk will be an expectant or new mother. The anticipated location of this equipment can be a PHC center. The important value addition is the reduction in the burden of data collection on front-line healthcare providers. The self-service will allow the users to sign up and remotely choose a health facility for appointment scheduling. Appointment scheduling is beyond the scope of this discussion and not the focus of this use case, but it requires the user to enroll through their mobile phones. For this use case, the users will visit the health facility having the kiosk, and while they are in queue, each will first go and enter all his or her bio details into the kiosk if not already done so for that health facility. Also, the user will provide the necessary vital signs by standing on, clipping, or using the sensor interfaces, and the system will automatically obtain these readings. Obtaining the readings directly from the IoHT parts rather than entering them will reduce errors and improve availability of such data. Many PHCs do not even have weighing scales and other vital-sign-measurement kits.

Users may require guidance to use the kiosk initially, but this will soon come natural to them. The unique identifier for each user will be generated by the system and benchmarked to some user characteristics, like the phone number and user age. This kiosk can successfully replace the registration point at health facilities. The kiosk will be on the same wireless network as the tablets used by the health workers in the PHC center. The tablet presents a different view to the frontline health workers at these PHCs. Taking the view for the prenatal nurses, for instance, the view will have an interface where the provider will review the history entered by the client at the kiosk and use that to make a decision and update the facility database. The user on the other end could also view his or her health information from the kiosk by checking on the information tab. For returning users, when they visit the kiosk, only their vital signs and biodata updates will be necessary.

For effectiveness and to ensure the system is operationally efficient, the system will be restricted to specific users at peak periods. For example, on prenatal clinic days, only pregnant women will be allowed to use the system for the duration of the prenatal service. This will ensure that the people who need it the most are the ones using the system. This system will make data available for decision making and high-quality vital signs. The system can even be extended such that a user can register symptoms at the health facility or to the cloud system through his or her mobile device. This feature of remote update is not the focus of this chapter, however.

4.4 Conclusion and Future Perspective

In this chapter, we looked at the current challenges health systems face globally and we reviewed how emerging technologies, like autonomous systems, have changed the face of medicine. One of the areas of interest was shifting the skilled work of clinical front-line health to workers while automating nonclinical tasks, such as registration and collection of some vital statistics, away from them so that they can focus on their responsibilities. The information revolution in the health sector was highlighted and the failed promises of digitization discussed. The opportunities for leveraging the many IoT and embedded devices together to address the issue of preventable deaths of new and expectant mothers in developing country were flagged.

Of the many use cases of IoHT, the chapter singled out designing a hypothetical maternal health self-service kiosk. The OSI architecture stack was used to model the kiosk using readily available components, elaborating on the power architecture, the network architecture, and the user flow strategy. In the future, this work will be prototyped to demonstrate feasibility. Also, the future of IoHT technology is to continue to expand as more embedded devices are made available and compatible with more medical and smart devices that allow users to manage and control their own health. These devices will hopefully be integrated within the IoHT network in the future but be based on the aggressive uptake of IoT in many healthcare use cases.

References

1. Michalak, J., Schreiner, N. J., Tennis, W., Szekely, L., Hale, M., Guastello, S. (2010). The patient will see you now. *Am. J. Nurs.*, **110**(1):61–63, doi:10.1097/01.NAJ.0000366060.37099.38.

2. Benson, T., Grieve, G. (2016). *Principles of Health Interoperability*, 3rd ed. Springer, doi:10.1007/978-3-319-30370-3.

3. WHO (2018). *World Health Statistics 2018: Monitoring the SDGs*.

4. Delgado, S. (2016). The digital doctor: hope, hype, and harm at the dawn of medicine's computer age. *Crit. Care Nurse*, **36**(4):84–85, doi:10.4037/ccn2016880.

5. Office of National Coordinator Health IT. https://www.healthit.gov. Accessed on 10 Oct. 2019.

6. Islam, S. M. R., Kwak, D., Kabir, M. H., Hossain, M., Kwak, K. S. (2015). The Internet of Things for health care: a comprehensive survey. *IEEE Access*, **3**:678–708, doi:10.1109/ACCESS.2015.2437951.

7. Schreier, G. (2014). The Internet of Things for personalized health. *Stud. Health Technol. Inform.*, **200**:22–31. doi:10.3233/978-1-61499-393-3-22.

8. da Costa, C. A., Pasluosta, C. F., Eskofier, B., da Silva, D. B., da Rosa Righi, R. (2018). Internet of Health Things: toward intelligent vital signs monitoring in hospital wards. *Artif. Intell. Med.*, **89**:61–69, doi:10.1016/j.artmed.2018.05.005.

9. Raad, M. W., Sheltami, T., Soliman, M. A., Alrashed, M. (2018). An RFID based activity of daily living for elderly with Alzheimer's. In *Internet of Things (IoT) Technologies for HealthCare*. HealthyIoT 2017, Ahmed, M., Begum, S., Fasquel, J. B., eds., Lecture Notes of the Institute for Computer Sciences, Social Informatics and Telecommunications Engineering, vol. 225. Springer, Cham, doi:10.1007/978-3-319-76213-5_8.

10. Paschou, M., Sakkopoulos, E., Sourla, E., Tsakalidis, A. (1970). Health Internet of Things: metrics and methods for efficient data transfer. *Simul. Model. Pract. Theory*, **34**:186–199, doi:10.1016/j.simpat.2012.08.002.

11. Guinard, D., Trifa, V. (2016). *Building the Web of Things: With Examples in Node. js and Raspberry Pi*. Manning Publications.

12. Moore, G. E. (1965). Cramming more components onto integrated circuits. *Electronics*, **38**(8):4.

13. Rodrigues, J. J. P. C., De Rezende Segundo, D. B., Junqueira, H. A., Sabino, M. H., Prince, R. M. I., Al-Muhtadi, J., De Albuquerque, V. H. C. (2018). Enabling technologies for the Internet of Health Things. *IEEE Access*, **6**:13129–13141, doi:10.1109/ACCESS.2017.2789329.

14. Miah, M. R., Hussain, M. S. (2018). A unique smart eye glass for visually impaired people. In *2018 International Conference on Advancement in Electrical and Electronic Engineering (ICAEEE)*, Gazipur, Bangladesh, pp. 1–4, doi:10.1109/ICAEEE.2018.8643011.

15. Jagtap, N., Wadgaonkar, J., Bhole, K. (2016). Smart wrist watch. In *2016 IEEE Students' Conference on Electrical, Electronics and Computer Science (SCEECS)*, Bhopal, pp. 1–6, doi:10.1109/SCEECS.2016.7509273.

16. Farjadian, A. B., Sivak, M. L., Mavroidis, C. (2013). SQUID: sensorized shirt with smartphone interface for exercise monitoring and home rehabilitation. *2013 IEEE 13th International Conference on Rehabilitation Robotics (ICORR)*, pp. 1–6, doi:10.1109/ICORR.2013.6650451.

17. Wu, Y. C., Chen, P. F., Hu, Z. A., Chang, C. H., Lee, G. C., Yu, W. C. (2009). A mobile health monitoring system using RFID ring-type pulse sensor. *2009 Eighth IEEE International Conference on Dependable, Autonomic and Secure Computing*, Chengdu, pp. 317–322, doi:10.1109/DASC.2009.136.

18. De Pessemier, T., Martens, L. (2018). Heart rate monitoring, activity recognition, and recommendation for e-coaching. *Multimed. Tools Appl.*, **77**:23317–23334, doi:10.1007/s11042-018-5640-2.

19. Simonnet, M., Gourvennec, B., Billot, R. (2016). Connected heart rate sensors to monitor sleep quality: electrodes, chest belt and smartwatch users acceptability. *2016 IEEE First International Conference on Connected Health: Applications, Systems and Engineering Technologies (CHASE)*, Washington, DC, pp. 344–345, doi:10.1109/CHASE.2016.38.

20. Bugeja, S., Garg, L., Audu, E. E. (2016). A novel method of EEG data acquisition, feature extraction and feature space creation for early detection of epileptic seizures. *2016 38th Annual International Conference of the IEEE Engineering in Medicine and Biology Society (EMBC)*, Orlando, FL, pp. 837–840, doi:10.1109/EMBC.2016.7590831.

21. Al-Odat, Z. A., Srinivasan, S. K., Al-qtiemat, E., Dubasi, M. A. L., Shuja, S. (2018). IoT-based secure embedded scheme for insulin pump data acquisition and monitoring. In *CYBER 2018: The Third International Conference on Cyber-Technologies and Cyber-Systems*, pp. 90–93.

22. Istepanian, R. S. H., Hu, S., Philip, N. Y., Sungoor, A. (2011). The potential of Internet of m-health Things m-IoT for non-invasive glucose level sensing. *Conf. Proc. IEEE Eng. Med. Biol. Soc.*, **2011**:5264–5266, doi:10.1109/IEMBS.2011.6091302.

23. Dohr, A., Modre-Osprian, R., Drobics, M., Hayn, D., Schreier, G. (2010). The Internet of Things for ambient assisted living. In *2010 Seventh International Conference on Information Technology: New Generations*, Las Vegas, NV, pp. 804–809, doi:10.1109/ITNG.2010.104.

24. Yang, G., Xie, L., Mäntysalo, M., Zhou, X., Pang, Z., Xu, L. Da, Kao-Walter, S., Chen, Q., Zheng, L. R. (2014). A Health-IoT platform based on the integration of intelligent packaging, unobtrusive bio-sensor, and intelligent medicine box. *IEEE Trans. Ind. Inf.*, **10**(4):2180–2191, doi:10.1109/TII.2014.2307795.

25. Yang, L., Ge, Y., Li, W., Rao, W., Shen, W. (2014). A home mobile healthcare system for wheelchair users. In *Proceedings of the 2014 IEEE 18th International Conference on Computer Supported Cooperative Work in Design (CSCWD)*, Hsinchu, pp. 609–614, doi:10.1109/CSCWD.2014.6846914.

26. Rasid, M. F. A., Musa, W. M. W., Kadir, N. A. A., Noor, A. M., Touati, F., Mehmood, W., Khriji, L., Al-Busaidi, A., Ben Mnaouer, A. (2014). Embedded gateway services for Internet of Things applications in ubiquitous healthcare. In *2014 2nd International Conference on Information and Communication Technology (ICoICT)*, Bandung, pp. 145–148, doi:10.1109/ICoICT.2014.6914055.

27. Jara, A. J., Zamora-Izquierdo, M. A., Skarmeta, A. F. (2013). Interconnection framework for mHealth and remote monitoring based on the Internet of Things. *IEEE J. Sel. Areas Commun.*, **31**(9):47–65, doi:10.1109/JSAC.2013.SUP.0513005.

28. Federal Ministry of Health (2016). Nigeria national eHealth strategy 2015–2020, Abuja, Nigeria.

29. DPPI–MoHS (2018). National digital health strategy (2018–2023), Sierra Leone.

30. Mittelstadt, B. (2017). Ethics of the health-related Internet of Things: a narrative review. *Ethics Inf. Technol.*, **19**:157–175, doi:10.1007/s10676-017-9426-4.

31. Shahamabadi, M. S., Ali, B. B. M., Varahram, P., Jara, A. J. (2013). A network mobility solution based on 6LoWPAN hospital wireless sensor network (NEMO-HWSN). In *2013 Seventh International Conference on Innovative Mobile and Internet Services in Ubiquitous Computing*, Taichung, pp. 433–438, doi:10.1109/IMIS.2013.157.

32. Laranjo, I., Macedo, J., Santos, A. (2013). Internet of Things for medication control: E-health architecture and service implementation. *Int. J. Reliab. Qual. E-Healthcare*, **2**(3):1–15, doi:10.4018/ijrqeh.2013070101.

33. Jara, A. J., Belchi, F. J., Alcolea, A. F., Santa, J., Zamora-Izquierdo, M. A., Gómez-Skarmeta, A. F. (2010). A pharmaceutical intelligent information system to detect allergies and adverse drugs reactions based on Internet of Things. In *2010 8th IEEE International Conference on Pervasive Computing and Communications Workshops (PERCOM Workshops)*, Mannheim, pp. 809–812, doi:10.1109/PERCOMW.2010.5470547.

34. Pang, Z., Tian, J., Chen, Q. (2014). Intelligent packaging and intelligent medicine box for medication management towards the Internet-of-Things. In *16th International Conference on Advanced Communication Technology*, Pyeongchang, pp. 352–360, doi:10.1109/ICACT.2014.6779193.

35. Vicini, S., Bellini, S., Rosi, A., Sanna, A. (2012). An Internet of Things enabled interactive totem for children in a living lab setting. In *2012 18th International ICE Conference on Engineering, Technology and Innovation*, Munich, pp. 1–10, doi:10.1109/ICE.2012.6297713.

36. Vazquez-Briseno, M., Navarro-Cota, C., Nieto-Hipolito, J. I., Jimenez-Garcia, E., Sanchez-Lopez, J. D. (2012). A proposal for using the Internet of Things concept to increase children's health awareness. In *CONIELECOMP 2012, 22nd International Conference on Electrical Communications and Computers*, Cholula, Puebla, pp. 168–172, doi:10.1109/CONIELECOMP.2012.6189903.

37. Chung, W. Y., Lee, Y. D., Jung, S. J. (2008). A wireless sensor network compatible wearable U-healthcare monitoring system using integrated ECG, accelerometer and SpO$_2$. In *2008 30th Annual International Conference of the IEEE Engineering in Medicine and Biology Society*, Vancouver, BC, pp. 1529–1532, doi:10.1109/iembs.2008.4649460.

38. Sebestyen, G., Hangan, A., Oniga, S., Gal, Z. (2014). eHealth solutions in the context of Internet of Things. In *2014 IEEE International Conference on Automation, Quality and Testing, Robotics*, Cluj-Napoca, pp. 1–6, doi:10.1109/AQTR.2014.6857876.

39. Bazzani, M., Conzon, D., Scalera, A., Spirito, M. A., Trainito, C. I. (2012). Enabling the IoT paradigm in E-health solutions through the VIRTUS middleware. In *2012 IEEE 11th International Conference on Trust, Security and Privacy in Computing and Communications*, Liverpool, pp. 1954–1959, doi:10.1109/TrustCom.2012.144.

40. Guinard, D. (2017). Workshop: building the Web of Things. CTO-co-founder.

41. Google Developers (2019). Save data using SQlite. In *Android Doc.* https://developer.android.com/training/data-storage/sqlite, accessed on 18 Oct 2019.

42. TinyOS (2013).

43. Upton, E., Halfacree, G. (2016). *Raspberry Pi® User Guide.* doi:10.1002/9781119415572.

44. Fathah, A. (2013). Design of a boost converter. Retrieved from http//ethesis.nitrkl.ac.in/4811/1/109EE0612.pdf.

45. Yoffie, D. B., Kim, R. (2009). E ink in 2008. HBS Case Collection November 2008 (revised April 2009).

46. Node.js Foundation (2015). https://nodejs.org/en/foundation/, accessed on 16 Oct 2019.

Chapter 5

Internet of Things for Smart Healthcare and Digital Well-Being

Niloy Sarkar[a] and Amitava Das[b]

[a] The Neotia University, West Bengal, India
[b] NSHM Knowledge Campus, Durgapur, India
dr.niloysarkar@gmail.com, amitavasdas@gmail.com

Sensing and communication devices and corresponding software have become resourceful for "Information Technology (IT)" keys in diverse frameworks. The promising technology known as the Internet of Things has enabled both experts and investigators to design innovative solutions in smart healthcare. There is extensive exploration of IoT-based healthcare, mainly because of its valuable inferences, including sophisticated quality, lower cost of services, and consistent care. This chapter explores IoT-based healthcare, including its applications and enabling technologies, as well as critical challenges.

5.1 Introduction

The advancement in information and communication technologies has opened the door for innovations in many aspects of daily life. See Fig. 5.1 for what the future may look like with the promising

Artificial Intelligence and the Fourth Industrial Revolution
Edited by Utpal Chakraborty, Amit Banerjee, Jayanta Kumar Saha, Niloy Sarkar, and Chinmay Chakraborty
Copyright © 2022 Jenny Stanford Publishing Pte. Ltd.
ISBN 978-981-4800-79-2 (Paperback), 978-1-003-15974-2 (eBook)
www.jennystanford.com

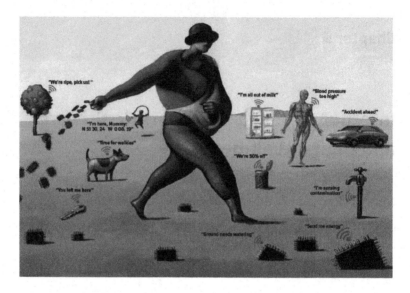

Figure 5.1   Vision of Internet of Things (authorized by Jon Berkeley).

technology identified as the Internet of Things (IoT) becoming a way of life. IoT has pushed practitioners and researchers equally to project advanced solutions in different contexts, specifically in healthcare. IoT devices have many applications in healthcare, accounting for over 30% of its application in all fields [1].

Rapid advancements in cloud computing, mobile applications, and wearable devices have facilitated IoT's role in transforming the traditional approach to healthcare into smart and personalized healthcare. IoT-enabled healthcare systems monitor several medical parameters, such as blood pressure and glucose levels, as well as body temperature, using smart sensors, computer networks, and a remote server. These healthcare systems could also provide basic suggestions for treatment, which is helpful, particularly when applied in homes or communities [2, 3].

As noted, IoT-assisted healthcare investigation is important because of its significance, counting advanced class and minor charge of amenities and dependable protective care. IoT-based healthcare is an active interdisciplinary research area where different methodologies, ranging from those of engineering and computer science to those of behavioral science, are deployed to design innovative information technology (IT) solutions to practical medical issues. Further research is needed to provide an overview of IoT—its architectures and various technologies—in the healthcare context. Therefore, this chapter aims to review the current literature on IoT-enabled healthcare. It illustrates the application of IoT in healthcare, as well as its enabling technologies and critical challenges.

IoT has garnered a lot of attention in recent years because of the possibilities it provides for relieving the stress on healthcare systems caused by an ageing population and a rise in chronic illness. Regulation is a key issue restricting progress in this area. This chapter describes a classic approach for application of IoT in healthcare and well-being systems. This chapter then presents the state-of-the-art research relating to each area, evaluating its strengths, weaknesses, and overall suitability for IoT-assisted healthcare systems. The challenges that the healthcare IoT faces, such as security, privacy, wearability, and low-power operation, are presented and referenced for future research directions.

Progress of human health and well-being is the vital objective of any economic, technological, and social development. The rapid aging of the population is one of the macropowers that will transform the world dramatically. Therefore, the applications of IoT technologies for in-home healthcare have been naturally highlighted in the strategic research roadmaps.

This chapter delivers a comprehensive study of the state-of-the-art technologies that fall within smart healthcare. Emphasis is placed on sensors for monitoring various health parameters, short- and long-range communications standards, and cloud technologies. This chapter differentiates itself from the previous major survey contributions by considering every essential component of an IoT-based smart healthcare system.

5.2 Internet of Things

The word "Internet of Things" was first used in 1999 by British technology pioneer Kevin Ashton to define a system in which objects in the physical world could be linked to the Internet by sensors. Ashton devised the term to illuminate the power of connecting radio-frequency identification (RFID) tags used in corporate supply chains to the Internet in order to count and track goods without the need for human involvement.

In 2003, the auto-ID center released the electronic product code (EPC) network. The EPC helps track objects moving from one location to another. This gives an idea of IoT implementation. RFID implementation further cements the opportunities for developing IoT as a new IT paradigm in both academic and industrial environments. At present, IoT has come to be a popular term for describing setups in which Internet connectivity and computing competency extend to a variety of objects, devices, sensors, and daily use items.

While the term "Internet of Things" is relatively new, the concept of combining computers and networks to monitor and control devices has been around for decades.

By the late 1970s, for example, systems for remotely monitoring meters on the electrical grid via telephone lines were already in commercial use. In the 1990s, advances in wireless technology allowed machine-to-machine (M2M) enterprise and industrial solutions for equipment monitoring and operation to become widespread. Many of these early M2M solutions, however, were based on closed purpose-built networks and proprietary or industry-specific standards, rather than on Internet Protocol–based networks and Internet standards.

The popular World Wide Web has become almost the same as the Internet itself. Web technologies expedite most interactions between people and content, making it a defining characteristic of the present Internet experience. The Web-based skill is largely characterized by the active engagement of users downloading and generating content through computers and smartphones. If the growth forecasts about IoT become a reality, we may see a shift toward more passive Internet interaction by users, with objects such as car components, home appliances, and self-monitoring devices; these devices will send and receive data on the user's behalf, with little human involvement or even cognizance.

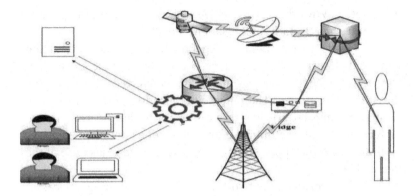

Figure 5.2 Classic Internet of Things (IoT) communication.

The concept of Internet working in IoT makes human life much easier than before, which is why IoT has gained our attention. Moreover, IoT is the most promising solution for the healthcare industry because it helps patients to manage their own diseases and receive help in most emergency cases via their mobiles. It is anticipated that the demand for personal healthcare applications will increase sharply. In the traditional healthcare model, the quality and scale of medical service cannot meet the needs of patients [4]. Therefore, it is of great significance to establish a family-oriented remote medical surveillance system based on mobile Internet. Generally, the provision of healthcare facilities through mobile devices is called m-health, where mobile devices are used to store, analyze, and transmit health statistics captured from multiple resources, including sensors and other biomedical acquisition systems [5, 8, 17].

Figure 5.2 shows how IoT communicates with other network devices. Doctors, patients, and the rest of the networking system are connected to each other. All record is digital and saved in databases that doctors and other clinical staff can access.

By an m-health service, we can easily provide quality medical services and medication as per patient needs [4]. IoT-based systems are responsible for the full care of the patient and are flexible because they can be set as per the patient's conditions and parameters. With this approach, we will be sure about the present and future health status of the patient [26, 28].

IoT may force a shift in thinking if the most common interaction with the Internet and the data derived from that interaction comes from passive engagement with connected objects in the broader environment. The potential realization of this outcome— a "hyper connected world"—is a testament to the general-purpose nature of the Internet architecture, which does not place inherent limitations on the applications or services that can make use of the technology.

IoT is a worldwide network of intercommunicating devices. It integrates ubiquitous communications, pervasive computing, and ambient intelligence. IoT is a vision where "things," especially everyday objects, such as all home appliances, furniture, clothes, vehicles, roads, and smart materials, are readable, recognizable, locatable, addressable, or controllable via the Internet. This will provide the basis for many new applications, such as energy monitoring, transport safety systems, and building security [18, 20]. This vision will surely change with time, especially as synergies

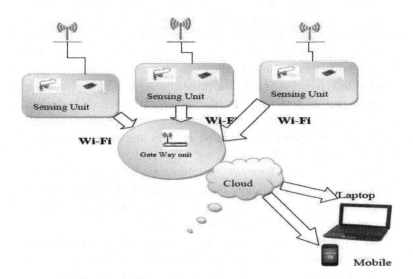

Figure 5.3 Overview of a wireless sensor network.

between identification technologies, wireless sensor networks, intelligent devices, and nanotechnology will enable a number of advanced applications. Innovative use of technologies such as RFID, near-field communication, ZigBee, and Bluetooth, is contributing to creating a value proposition for patrons of IoT [6, 7].

IoT will connect the world's objects in a sensory as well as intelligent manner by combining technological developments in item identification, sensors and wireless sensor networks (Fig. 5.3), embedded systems, and nanotechnology.

In this chapter, we will discuss mainly the applications, benefits, and future challenges of IoT based on the work done by different researchers in the field of IoT. The main aim of this chapter is to provide an overall idea of what IoT is, the different forms of applications it has adopted, and how it is providing a solution for the problems faced by the global healthcare industry [9, 10].

5.3 Technologies behind IoT

IoT systems in healthcare can operate in different scenarios and accordingly make effective decisions with the help of communication and sensory devices and the applicable software.

5.3.1 Cloud Computing

Cloud computing is a public term in today's IT world. The most appropriate definition of cloud computing is provided by Borko Furht of Florida Atlantic University, who explains it as "a new style of computing in which dynamically scalable and often virtualized resources are provided as a service over the Internet." The amalgamation of IoT and cloud computing is of great significance. Cloud computing, powerful storage, processing, and serviceability, combined with IoT's ability of information collection, compose a real network between people and items and among the items themselves [21, 23].

Cloud computing is a computing standard where a large pool of systems is connected on private or public networks to provide dynamically scalable infrastructure for application, data, and file storage. With the advent of this technology, the cost of computation, application hosting, content storage, and delivery is reduced meaningfully.

Cloud computing is a practical approach to understanding direct cost benefits, and it has the potential to transform a data center from a capital-intensive setup to a variable-priced environment.

The idea of cloud computing is based on a very fundamental principle of reusability of IT capabilities [16]. The difference that cloud computing brings compared to the traditional concepts of "grid computing," "distributed computing," "utility computing," and "autonomic computing" is to broaden horizons across organizational boundaries. Despite differences in defining cloud computing, there is at least one common understanding, that a layered architecture exists. The number of layers is not fixed, and the foundation is versatile hardware. The lowest layer is the hardware, on top of which comes the software platform, on top of which is built the software layer [22, 24].

5.3.2 Sensors

Sensors form the heart of IoT-based systems as they are on-the-ground devices performing the critical part of monitoring processes, taking measurements, and collecting data. With advancements in sensing technologies, it is possible to obtain continuous data from objects or even from living creatures, such as humans. For example, the pulse oximeter that was invented in the 1970s is used as a major device for diagnosis. It helps a physician monitor a patient's heart rate and blood oxygen saturation (SpO_2), which are critical for emergency services [25, 27]. Other types of sensors are temperature, pressure, water quality, and smoke sensors. The instrument that analyzes motion is a complicated device that comprises various sensors, such as accelerometers, gyroscopes, and surface electrodes. It is possible to transform all the received data from the sensors into a digital form and instantly transmit it over a network. The popularity of wireless sensors has made it thinkable for people to wear portable sensors capable of automated data collection and transfer.

5.3.3 Location

In modern tracking systems, real-time location systems (RTLS) help locate objects. The global positioning system (GPS) is considered the most important RTLS. This satellite-dependent navigation system is capable of locating objects under various weather conditions. For healthcare applications, GPS can help in precisely locating ambulances, patients, doctors, and so on. However, these navigation systems (e.g., the GPS or China's Beidou system [BDS]) are ineffective indoors due to the construction structure, which hampers the satellite's transmission signals. Hence, it becomes necessary to develop a local positioning system (LPS) that can replace the GPS and work with better accuracy indoors [17]. An LPS measures the radio signals that travel between an object and an array of receivers aimed to locate it. There is a great possibility to create smart indoor positioning network systems by combining the GPS or the BDS with an LPS in a high-bandwidth wireless communication network.

5.3.4 Communication

When considering an IoT-based system, communication technologies enable network infrastructure. IoT networks are based on heterogeneous frequencies, standards, and transmission rates for transferring data. These networks can be further classified as long-distance and short-distance technologies. Long-distance technologies are intended to deal with regular means of communication, such as the Internet and mobile phones. Short-distance communication mostly utilizes wireless technologies, such as Bluetooth, Infrared Data Association, Wi-Fi, ultrawideband, and RFID [29, 31, 32]. All these technologies enable data transmission over a short distance. These technologies may differ in terms of the installation cost, transmission rates, distance, the number of entities, power consumption, and maintenance cost, etc., according to the differences in working radio frequencies and security standards.

5.3.5 Identification

A working model of IoT may have a large number of nodes. An authorized node may generate data and could have access to data, irrespective of the data's location. To achieve this goal, it becomes necessary to efficiently seek and identify the nodes. The identification process assigns a unique identification number (UID) to each node in order to enable unambiguous information exchange via the node. Each individual resource in a system is given a digital UID. For example, in a hospital, a doctor, a nurse, or any other staff member is provided with a UID. This helps create relations among different entities in the digital domain. It enables the prompt location of the available objects in the network without fail. Various standards have been proposed for the unique identification of objects in the digital domain, such as a universally unique identifier. One such standard has been developed by the Open Software Foundation [30].

5.4 Healthcare and the Internet of Things

The healthcare industry is developing as the largest area of interest globally, but with the advancement of this industry, the cost for healthcare is spinning out of control of the common people, especially in those countries where there is no social assurance.

If we analyze the reasons behind this escalating cost year after year, the foremost contributing factor would be the cost of new and advanced medical equipment and the fourth- and fifth-generation medicines, apart from factors like infrastructure cost; quality cost; maintenance cost; and obviously the cost of skilled manpower, like doctors, nurses, and paramedics.

Machine learning and IoT would be very beneficial in minimizing the cost of healthcare. Apart from reducing cost, machine and

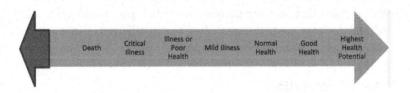

Figure 5.4 The health continuum.

IoT offer several other benefits in the healthcare sector. In recent research, it has been found that breast cancer in women can be deduced at a very early stage, even up to five years ahead of the actual incidence [33].

Massachusetts Institute of Technology (MIT) Computer Science & Artificial Intelligence Laboratory has developed a new dip learning–based AI prediction model that can anticipate the inception of breast cancer even more than five years in advance. Black women are 42% more likely than white women to die from breast cancer, and the new AI device (MIT dip learning tool) is specially trained on mammograms, specifically for black women, who are more vulnerable to breast cancer, giving marvelous results [34]. MIT is also hopeful that this device can be used to improve the deduction of other diseases that have similar problems with existing risk models. IoT-based solutions have been readily accepted by the healthcare industry, which is one of the fastest growing among all sectors.

5.5 Internet of Things for Health

IoT for health is basically an IoT-based solution that includes a network architecture that allows the connection between a patient and healthcare facilities, for example, an IoT-based e-health system for electrocardiography [11], heart rate [12], electroencephalogram [13], diabetes [14], and other different kinds of monitors of the vital signs of the body based on biomedical sensors (Fig. 5.5). It includes pulse, SpO_2, airflow (breathing), body temperature, glucometer, galvanic skin response, blood pressure, patient position (accelerometer), and electromyography [15, 19]. Data from patients can be collected through sensors and processed by applications developed for a user terminal, such as computers, smartphones, smartwatches, or even specific embedded devices [4].

The Internet of Health Things (IoHT) can support many medical areas, like care for pediatric and geriatric patients, supervision of chronic diseases, and management of personal health and fitness.

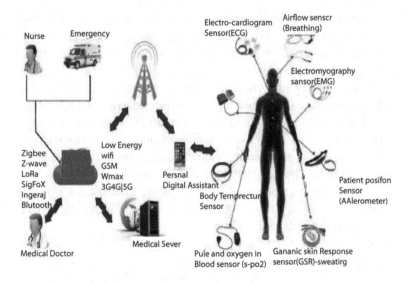

Figure 5.5 Remote healthcare monitoring system.

5.5.1 Digital Wellness

Wellness describes a range of health factors, from the highest potential of health (optimum level) to the onset of illness or the beginning of a maladaptive stage (Fig. 5.4).

Therefore, continuous monitoring is the best possible solution toward digital wellness, at least for those who fall in the vulnerable category.

5.5.2 Continuous Health Monitoring

Faster medical treatment saves lives. Continuous health monitoring at an individual level for over a period would be an expensive method of maintaining good health, but application of IoT can bring this within the reach of common people, and one such example is TREWS.

5.5.3 Easy Way of Continuous TREWS

It's one of the very useful devices developed by the Machine Learning and Health Lab of Johns Hopkins University. The targeted

Figure 5.6 Smartphone for the healthcare system.

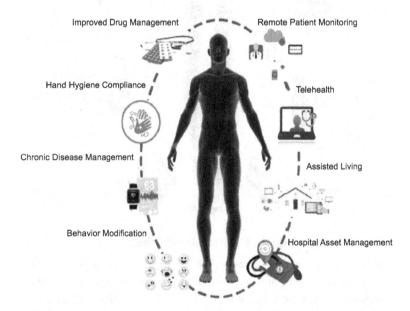

Figure 5.7 Ambient assisted living system.

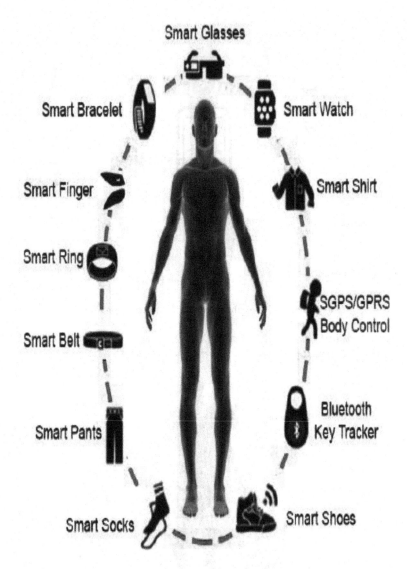

Figure 5.8 Wearable technology.

real-time early warning system (TREWS) can detect the symptoms of any health emergency at least 12 to 24 hours before a doctor can. Studies show that there is a 60% increase in the performance of disease detection (limited to the nature of the disease) after the application of TREWS.

Annually, almost 750,000 people get affected terminally by blood infection, commonly known as sepsis, and it is a medical emergency. The most vulnerable groups are older adults and children. Johns Hopkins claims that the application of TREWS has helped minimize mortality and morbidity among these groups.

5.6 Integration of Different Disciplines of Science toward Better Application of AI and IOT

Integration of different disciplines of science is essential for inventing various applications of AI in the vast domain of the healthcare field.

Better implementation of AI may require interdisciplinary collaboration of different fields of study (Fig. 5.9); a symbiotic relationship between the experts of different fields could be the key behind this integration. Newer fields of studies are likely to be introduced to take care of the developments and to improve this collaboration.

5.7 AI: Major Areas of Application in the Health Field

Healthcare is a complex field of study. Integration of separate components and their utilization is fundamental to the success of this field. To improve the care provided to patients, AI, at its initial level, has already been applied in the following fields:

- Helping in the diagnostic process
- Developing treatment protocols
- Developing drugs
- Monitoring patients
- Helping in personalized medication
- Managing data, including medical records
- Providing digital consultation
- Identifying precision medicine
- Helping in precision surgery

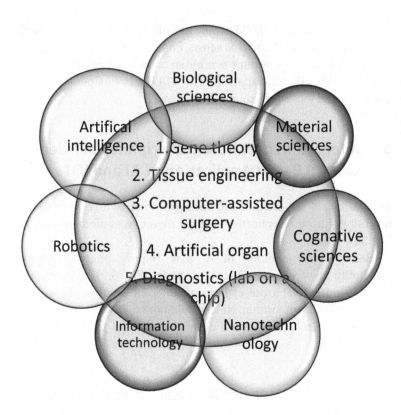

Figure 5.9 Sectoral integration for better AI applications.

- Doing repetitive jobs

5.8 Present Applications of AI in Healthcare

- AiCure: Developed by the National Institute of Health and applied for medication management for individual patients.
- Molly: Developed by the startup Sense.ly. It's a digital nurse, very effective for at-home care in the case of chronically ill patients.
- Amazon Alexa: Developed by Amazon and used at the Boston Children's Hospital. It gives basic health information and advice to guardians of ill children. The application answers all questions on medications and symptoms.

- TREWS: Developed by Machine Learning and Healthcare Lab, Johns Hopkins University, for vital data assimilation and early warning, especially for cases of sepsis.
- iDOCTOR: Developed by IBM. It helps in data assimilation and analysis. It has application in health system analysis.

AI-Rad Companion Chest CT: Developed by Siemens Healthineers. It is made for highlighting, quantification, and measurement of defined anatomies and abnormalities and can automatically report the findings in a structured table that is sent to PACS.

The AI-Pathway Companion5: Developed by Siemens Healthineers. It helps in data integration and evaluation. It also provides aggregated patient history and findings, projections on the patient's status in the clinical pathway following clinical guidelines, and suggestions for the next steps to follow.

5.9 Conclusion and Future Work

The rapid advancements in cloud computing, mobile applications, and wearable devices facilitate IoT's role in transforming the traditional approach to healthcare into smart and personalized healthcare. IoT-enabled healthcare systems can be categorized into three primary areas: monitoring, controlling, and information sharing.

Despite the value of implementing IoT solutions in healthcare, it might encounter some challenges, for example, in terms of security, scalability, and mobility. Security and privacy remain the most critical issues. The security of IoT-healthcare building blocks should be prioritized, including that of medical sensors, the network of IoT nodes, and cloud services. Additionally, proper policies and technical security measures are essential to enable data sharing among authorized devices, users, and organizations. Due to the potential risk of storing sensitive health data in IoT cloud services, further research should consider designing data-transparent cloud services. Researchers can employ a design science approach to develop IoT cloud services in a way that healthcare data would be traced and its usage controlled.

Nevertheless, this review does not provide a deep understanding of some fundamental topics, including IoT topologies, architectures, and platforms, in addition to the security requirements and challenges in the proposed model. There are other sets of technologies that remain unexplored in this chapter, such as big data, augmented reality, and cognitive systems, and could be explored further in future studies.

Finally, policies and regulations are very important in the healthcare sector and should be considered in the near future as a new domain of IoT application.

To accomplish this goal, the most recent IoHT publications and products were identified, described, and analyzed. To conclude, there are many services and applications of IoHT that meet the societal needs but have still not been included in the mainstream of IoHT. The discussion in this chapter may help developers and entrepreneurs to build solutions that could be embraced by all parts of the society.

References

1. Van Den Broek, G., Cavallo, F., Wehrmann, C. (2010). *AALLIANCE Ambient Assisted Living Roadmap* (Ambient Intelligence and Smart Environments, vol. 6). IOS Press, Amsterdam, the Netherlands, p. 136.
2. Perrier, E. (2015). Positive Disruption: Healthcare, Ageing & Participation in the Age of Technology. The McKell Institute, Australia.
3. Gope, P., Hwang, T. (2016). BSN-care: a secure IoT based modern healthcare system using body sensor network. *IEEE Sens. J.*, **16**(5):1368–1376.
4. Zhu, N., Diethe, T., Camplani, M., Tao, L., Burrows, A., Twomey, N., Kaleshi, D., Mirmehdi, M., Flach, P., Craddock, I. (2015). Bridging e-health and the Internet of Things: the SPHERE project. *IEEE Intell. Syst.*, **30**(4):39–46.
5. Chang, S. H., Chiang, R. D., Wu, S. J., Chang, W. T. (2016). A context-aware, interactive M-health system for diabetics. *IT Prof.*, **18**(3):14–22.
6. Pasluosta, C. F., Gassner, H., Winkler, J., Klucken, J., Eskofier, B. M. (2015). An emerging era in the management of Parkinson's disease: wearable technologies and the Internet of Things. *IEEE J. Biomed. Health Inf.*, **19**(6):1873–1881.
7. Fan, Y. J., Yin, Y. H., Xu, L. D., Zeng, Y., Wu, F. (2014). IoT based smart rehabilitation system. *IEEE Trans. Ind. Inf.*, **10**(2):1568–1577.

8. Sarkar, S., Misra, S. (2016). From micro to nano: the evolution of wireless sensor-based health care. *IEEE Pulse*, **7**(1):21–25.

9. YIN, Y., Zeng, Y., Chen, X., Fan, Y. (2016). The Internet of Things in healthcare: an overview. *J. Ind. Inf. Integr.*, **1**:3–13.

10. Islam, S. M. R., Kwak, D., Kabir, H., Hossain, M., Kwak, K.-S. (2015). The Internet of Things for health care: a comprehensive survey. *IEEE Access*, **3**:678–708.

11. Dimitrov, D. V. (2016). Medical Internet of Things and big data in healthcare. *Healthcare Inf. Res.*, **22**(3):156–163.

12. Poon, C. C. Y., Lo, B. P. L., Yuce, M. R., Alomainy, A., Hao, Y. (2015). Body sensor networks: in the era of big data and beyond. *IEEE Rev. Biomed. Eng.*, **8**:4–16.

13. Laplante, P. A., Laplante, N. (2016). The Internet of Things in healthcare: potential applications and challenges. *IT Prof.*, **18**:2–4.

14. Blake, M. B. (2015). An Internet of Things for healthcare. *IEEE Internet Comput.*, **19**:4–6.

15. Po Yang, P., Amft, O., Gao, Y., Xu, L. (2016). Special issue on the Internet of Things (IoT): informatics methods for IoT-enabled health care. *J. Biomed. Inf.*, **63**:404–405.

16. Yin, Y., Zeng, Y., Chen, X., Fan, Y. (2016). The Internet of Things in healthcare: an overview. *J. Ind. Inf. Integr.*, **1**:3–13.

17. Benny, P. L. L., Henry, I., Guang-Zhong, Y. (2016). Transforming health care: body sensor networks, wearables, and the Internet of Things. *IEEE Pulse*, **7**:4–8.

18. Madakam, S., Ramaswamy, R., Tripathi, S. (2015). Internet of Things (IoT): a literature review. *J. Comput. Commun.*, **3**:164–173.

19. Ashton, K. (2009). That "Internet of Things" Thing. *RFiD J.*, **22**:97–114.

20. Gershenfeld, N., Krikorian, R., Cohen, D. (2004). The Internet of Things. *Sci. Am.*, **291**:76–81.

21. Tarouco, L. M. R., Bertholdo, L. M., Granville, L. Z., Arbiza, L. M. R., Carbone, F., Marotta, M., De Santanna, J. J. C. (2012). Internet of Things in healthcare: interoperability and security issues. In *IEEE International Conference Communications*, Sydney, Australia.

22. Mattern, F., Floerkemeier, C. (2010). From the Internet of Computers to the Internet of Things. In *From Active Data Management to Event-Based Systems*, Vol. 6462, Sachs, K., Petrov, I., Guerrero, P., eds. Springer, Berlin, Heidelberg, pp. 242–259.

23. Lee, I., Lee, K. (2015). The Internet of Things (IoT): applications, investments and challenges for enterprises. *Bus. Horiz.*, **58**:431–440.

24. Williams, P. A., Woodward, A. J. (2015). Cybersecurity vulnerabilities in medical devices: a complex environment and multifaceted problem. *Med. Devices (Auckland)*, **8**:305–316.

25. Miorandi, D., Sicari, S., De Pellegrini, F., Chlamtac, I. (2012). Internet of Things: vision, applications and research challenges. *Ad Hoc Networks*, **10**:1497–1516.

26. Jian, Z., Zhan, W., Zhuang, M. (2012). Temperature measurement system and method based on home gateway. Patent CN 201110148247.

27. Yang, L., Ge, Y., Li, W., Rao, W., Shen, W. (2014). A home mobile healthcare system for wheelchair users. In *Proceedings of the 18th International Conference on Computer Supported Cooperative Work in Design*, Hsinchu, Taiwan.

28. Sivagami, S., Revathy, D., Nithyabharathi, L. (2016). Smart health care system implemented using IoT. *Int. J. Contemp. Res. Comput. Sci. Technol.*, **2**(3).

29. Datta, S. K., Bonnet, C., Gyrard, A., Costa, R. P. F. D., Boudaoud, K. (2015). Applying Internet of Things for personalized healthcare in smart homes. In *Proceedings of the 24th Wireless and Optical Communication Conference*, Taipei, Taiwan.

30. Samuel, R. E., Connolly, D. (2015). Internet of Things-based health monitoring and management domain-specific architecture pattern. *Issues Inf. Syst.*, **16**:58–63.

31. Rahmani, A.-M., Thanigaivelan, N. K., Gia, T. N., Granados, J., Negas, B., Liljeberg, P., Tenhunen, H. (2015). Smart e-health gateway: bringing intelligence to Internet-of-Things-based ubiquitous healthcare systems. In *Proceedings of the Annual IEEE Consumer Communications and Networking Conference*, NV, USA.

32. Agrawal, S., Vieira, D. (2013). A survey on Internet of Things. *Abakós, Belo Horizonte*, **1**:78–95.

33. Densen, P. (2011). Challenges and opportunities facing medical education. Retrieved from https://www.ncbi.nlm.nih.gov/pmc/articles.

34. AumuellerJoerg–Siemens Heaithineers: Health Europa. Retrieved from www.Healtheuropa.eu.

35. Murdoch, T. B., Detsky, A. S. (2013). The inevitable application of big data to healthcare. *JAMA*, **309**:1351–1352.

Chapter 6

Automated Chatbots for Autism Spectrum Disorder Using AI Assistance

Vamsidhar Enireddy,[a] Karthikeyan C.,[a] and Ramkumar J.[b]

[a] *Department of Computer Science and Engineering, Koneru Lakshmaiah Education Foundation (KLEF), K L Deemed to be University, Vaddeswaram, Guntur, Andhra Pradesh, India*
[b] *Department of Computer Science and Engineering, SRM Institute of Science and Technology, Kattankulathur, Chennai, Tamil Nadu, India*
enireddy.vamsidhar@gmail.com, ckarthik2k@gmail.com, ram.kumar537@gmail.com

Autism, or autism spectrum disorder (ASD), refers to a vast range of conditions portrayed by repetitive behaviors and difficulties with social abilities, speech, and nonverbal communication. There are a number of subtypes in ASD, most due to the genes specific to autism and also due to the environmental effect. The effect of autism varies across individuals due to spectrum disorder difficulty. Detection of autism at an early stage has shown positive results according to some studies done on the disorder. In the United States alone, 1 out of 59 kids shows signs of autism. In India, 1 in 100 children shows positive signs of autism. A major problem in India is that parents of such children don't come out in the open about this problem due to the fear that the society may hate the child, making the condition worse for the child because he or she does not get proper treatment

Artificial Intelligence and the Fourth Industrial Revolution
Edited by Utpal Chakraborty, Amit Banerjee, Jayanta Kumar Saha, Niloy Sarkar, and Chinmay Chakraborty
Copyright © 2022 Jenny Stanford Publishing Pte. Ltd.
ISBN 978-981-4800-79-2 (Paperback), 978-1-003-15974-2 (eBook)
www.jennystanford.com

and care. Parents of such children also require special training in order to understand the behavior of their children and deal with it.

Although they look normal, autistic patients display mental behavior that is different from that of normal people and they need special care. Use of information technology and machine learning techniques can help to bridge the gap between the patient, the parent, and the doctor. A children dashboard is created and deployed on a digital slate. The dashboard contains a camera and allows touch and 3D interaction and is used only for this purpose. It also contains a conversational chatbot that can deliver the child's psychosomatic involvement based on depression and anxiety in order to provide behavioral therapy using artificial intelligence. These chatbots are easy to access, discover emotions and needs, and are easy to customize. The chatbot presents adapted responses based on the patient's mood and using a decision tree model provides appropriate personalized responses. The chatbot model is automated, has a compassionate conversation style of demonstrating empathy, and also provides weekly charts defining the child's state of mind. Having access to most recent advances, like chatbots, provides a genuine necessary break to people who experience the ill effects of autism or other disorders demonized by the general public. As AI and machines do not have emotions or feelings, they help individuals share their actual sentiments when such individuals realize that these machines will support them unconditionally and not pass judgment on them. The negative side of the development of such AI-based choices is that it makes us think that regular methods of ensuring psychological well-being, such as human connection, are outdated.

6.1 Introduction to Autism

Autism spectrum disorder (ASD) is a neuroformative issue portrayed by shortages in social correspondence and social association, notwithstanding limited, redundant pattern of conduct [1]. It may also refer to a scope of conditions portrayed by some level of impeded social conduct, correspondence, and language; a restricted scope of interests and exercises; and display of practices like animosity, self-damage, and fits that are novel to the individual and done repetitively [2, 3]. ASDs start in infancy and will in general endure into immaturity and adulthood. As a rule, the conditions become obvious during the initial five years of life. Intercession during early adolescence is critical to ensure the ideal improvement and prosperity of an individual with ASD. Observation of child advancement as a feature of routine maternal and child medicinal services is recommended [4].

It is vital that, when identified, youngsters with ASD and their families are offered pertinent data, administrations, referrals, and viable help as indicated by their individual needs. There is no solution available for ASD [5]. Proof-based psychosocial intercessions—be that as it may, for example, conduct treatment and abilities-preparing programs for guardians and different parental figures—can diminish challenges in correspondence and social conduct, with a constructive impact on the individual's prosperity and personal satisfaction. The human services needs of individuals with ASD are perplexing and require a scope of coordinated administrations, including well-being advancement, care, recovery administrations, and joint effort of different divisions, for example, training, business, and social sectors [5]. Intercessions for individuals with ASD and other formative issues should be combined with more extensive activities for making their physical, social, and attitudinal conditions increasingly available, comprehensive, and steady. Assistive innovation, for example, augmentative and elective correspondence, can profit individuals with chemical imbalance by supporting and improving correspondence, advancing freedom, and expanding social interactions [6].

6.1.1 Need to Study Autism

In 2018, the Centers for Disease Control and Prevention discovered that 1 out of 59 youngsters in the United States alone is determined to have a chemical imbalance range issue (ASD) [7]. It was also estimated that 1 in 37 boys and 1 in 151 girls are suffering from ASD. Boys are more prone to a chemical imbalance compared with girls. The analyzation to check the child for mental imbalance should start as early as two years of age rather than, say, at the age of four, where we would have already wasted two valuable years where treatment could have been provided. It is found that 31% of kids with ASD have a scholarly inability when tested for intelligence quotient (IQ) values, where their IQ < 70; 44% of them have a better range (IQ > 85); and the remaining possess an IQ in the range of 71–85 [6]. Chemical imbalance influences all ethnic and financial gatherings. Minority gatherings will in general be analyzed later and less frequently. Early intercession offers the best chance to encourage sound advancement and convey benefits over the life expectancy. There is no medicinal intervention for mental imbalance.

It has been assessed that 33% of individuals with chemical imbalance are nonverbal and that in children, 31% with ASD have a scholarly handicap (IQ < 70), facing noteworthy difficulties in day-to-day work, and 25% are in the marginal range (IQ 71–85). In the case of mental imbalance or a tendency to meander or get jolted, nearly 50% of the children are found to suffer with the problem of autism. About 66% of youngsters with mental imbalance between the ages of 6 and 15 have been harassed, and almost 28% of eight-year-olds having autism perform self-damaging practices, like head hammering, arm biting, and skin scratching. The main cause of death among 14-year-olds or younger with autism is found to be drowning, along with other causes, like meandering and dashing around.

6.1.2 Identification Symptoms

Although ASD symptoms are known by the age of five years and above, they can be identified at an early stage at two years of age. This also gives us the option to start treatment at the early stage. The following symptoms can be taken as early signs to identify ASD.

The symptoms presented here are to be checked if the child or toddler does not perform them [8]. They are only some of the symptoms that are important, such as the child or toddler making an eye-to-eye association, for instance, looking when being looked at or smiling when being smiled at; answering to his or her name; visually following objects or gestures when shown; using other means of communication, like waving or making noises, to grab the attention of others; trying to make faces; imitating others' movements; and playing with others with happiness and interest.

If the baby is not responding to these things, then it is better to meet a doctor or a specialist to confirm ASD.

It is more ominous when a toddler shows the following symptoms, and such a situation requires instantaneous assessment by the doctor [8]: if by six months, the child is not giving huge smiles or any other happy expressions; if by nine months, the child does not display to-and-fro sharing of sounds, grins, or other outward gestures; if by the age of one year, the child does not start reacting to her or his name; if the child does not make verbal expressions by the age of sixteen months; and if by the age of two years, the child does not have significant expression of phrases (which may not include imitating). These observations play an important role in helping identify ASD and show the difference between a child with ASD and a healthy child.

6.1.3 Challenges Faced in a Community

Children with ASD do not show any interest in the happenings in their surroundings and do not get involved with others. They also do not know the way to interact with others; are not able to play or make friends with other children; do not like to be touched, be held by others, or play group games; and cannot imitate or utilize toys in imaginative ways. They also have trouble in understanding feelings and emotions or discussing them. They do not show any eagerness to listen to or converse with other persons and show no enthusiasm in sharing their accomplishments with others [8].

The most important thing is the communication of these youngsters with others. They lack simple social communication skills with other children. These children do not like to interact with others, even those of the same age, and they want to lead solitary lives, in a reality of their own, segregated from other children.

6.1.3.1 Challenges in verbal communication

Children with ASD speak in an abnormal manner, with sounds of different characteristics, for example, in a different beat or pitch, compared to others. They repeat similar words or expressions recursively, frequently without open expectation. They tend to respond to questions by simply repeating the questions. They use language erroneously (linguistic mistakes, wrong words) and allude to themselves in the third person. These individuals have difficulty communicating with others, even when they want to, and they tend not to react to simple directions, small statements, etc. They take what is said literally (missing out on nuances of humor, incongruity, and mockery) [9].

6.1.3.2 Challenges in nonverbal communication

Children with ASD lack the ability to coordinate with people. For example, when a person is talking with a child with ASD or stating something, the child will not maintain eye contact with that person, simply behaving as though he or she is not concerned with what is being said. The most disturbing aspect is that such children show an outward appearance different from that of normal children, for example, in terms of movements and speech. They behave as robots, showing minimal gestures. They show surprised reactions to sights, scents, surfaces, and sounds and a very low response to loud noises [8].

6.2 Behavior of Autistic Children

Some common symptoms are summarized here to identify ASD at an early age: rocking to and fro; spinning around; repeating finger movements continuously; slamming the head against objects; continuously looking at bright objects, like lights; continuously waving their fingers before their eyes or flapping their hands; using their nails to scratch themselves; and paying attention to spinning objects and wheel movements. These children tend to place toys in a line instead of playing with them. They commonly repeat the same word continuously.

6.2.1 Reasons for Autism

Up to this point, most researchers have accepted that mental imbalance is caused, for the most part, by hereditary elements. Be that as it may, pivotal new research demonstrates that natural elements may also be significant in the advancement of mental imbalance. Infants might be brought into the world with a hereditary propensity to mental imbalance, which is then activated by something in the outer world, either while the individual in question is still in the belly or at some point after birth [9, 10]. It's imperative to take note that the world, in this specific circumstance, implies anything outside the body. It's not restricted to things like contamination or poisons in the climate. Indeed, one of the most significant conditions of autism happens while they are in the prebirth condition [9, 10].

6.2.2 Other Reasons That Contribute to ASD

During pregnancy, women should not consume antidepressants, especially in the first three months, since these may cause side effects and can have some influence on the fetus. Women should take care that during pregnancy they consume enough folic acid and supplements so that the required levels can be maintained because deficiency in these may have an influence on the fetus. Couple age is also a reason for ASD. Giving birth to a child at a late age increases the chances of ASD in the child. Issues after birth, such as neonatal iron deficiency and a baby born with less weight, can also cause ASD [8].

6.2.3 Negligence during Pregnancy

The couple has to take care that the would-be mother is not exposed to any poison, pollutants (including heavy metals), and pesticides. More research on these prebirth hazards is required. However, in case you are pregnant or attempting to get pregnant, there is no harm in your taking steps to lessen your infant's chances of facing chemical imbalance [8].

6.2.4 Formative Screening Assessment

This test, also known as development screening, is to test the abilities of the children, including their learning abilities and essential skills, and to identify any delays or abnormalities during the test. The specialist has to talk with the parents and talk to and play with the child to infer the abilities of the child. These tests also help the specialist to find out whether the child is able to learn, speak to, and play with other kids. It is to be noted that parents should not delay in getting their children to take these tests [11].

Regular and timely check-up of a child by a doctor will help identify any formative deferrals and disabilities. The child should be taken for a check-up at 9, 18, 24, and 30 months to the specialist doctors. Further screening analysis may be required when kids are born prematurely, have lower birth weights than normal, have kin with ASD, or show habits related with ASD. Although there are check-ups after the birth of a child, it should be noted that these check-ups are for normal health parameters of the child and the normal pediatrician tests do not cover tests for ASD. If the doctor finds any issue, he or she will refer the child to the specialist for a deeper assessment of the child's behavior [11].

6.2.5 Exhaustive Diagnostic Evaluation

Only after the final screening assessment can it be decided whether the child needs tests for ASD. This escalated review may combine reviewing the child's lead and progression and talking to the guardians. Similarly, vision screening, genetic testing, neurological testing, and other restorative testing may be recommended. Every so often, the basic thought authority may suggest that the child, along with the family, visit an expert for further assessment and assurance. Professionals who can provide solutions to the problem of autism are known as developmental pediatricians (who have remarkable experience in adolescent improvement and catering to children with phenomenal needs), child neurologists, and child psychologist [11].

6.2.6 Associated Medical and Mental Health Conditions

The problem with autism is that it can influence the entire body, and one of the problems associated with it is attention deficient hyperactivity disorder, which influences 30% to 60% of youngsters with chemical imbalance. It is estimated that more than half of the youngsters with chemical imbalance have at least one restlessness issue. Anxiety issues also influence 11% to 40% of youngsters and adolescents with chemical imbalance. Most of the findings suggest 7% of kids and 26% of grown-ups suffer from depression [6, 12]. Children with a mental imbalance are bound to experience the ill effects of at least one incessant recurrent gastrointestinal issue. Nearly 33% of individuals with a mental imbalance have epilepsy (seizure issue). Studies propose that schizophrenia influences somewhere in the range of 4% to 5% of grown-ups with a mental imbalance. On the other hand, schizophrenia influences about 1.1% of the all-inclusive community. Autism-related medical issues can have an effect over the life expectancy—from small kids to seniors. Almost a third (32%) of two-year-olds and older people with a mental imbalance are overweight and 16% are large. On the other hand, about 23% of two-year-olds and older people in the overall public are overweight and just 10% are therapeutically fat.

The medicines for these children should be administered properly. Only a few medicines are available and approved by the Food and Drug Administration, such as risperidone and aripiprazole, for treating autism-related disturbances and crabbiness.

6.3 Financial Burden on the Families and Effect on the Economy of a Country

The expense of research on ASD and treatment of Americans with a chemical imbalance came to US$268 billion in 2015 and would ascend to US$461 billion by 2025 without progressive compelling intercessions and backing over the life expectancy. Most of the expenses to research and treat mental imbalance in the United States are for grown-ups ($175 billion to $196 billion per year), in contrast to $61 billion to $66 billion every year for youngsters. By and large, medicinal uses for kids and teenagers with ASD were 4.1–6.2 times more than for those without a chemical imbalance. A section of the 2014 Achieving a Better Life Experience Act permits charge favored bank accounts for individuals with handicaps, including chemical imbalance, to be set up by states [6, 13].

Table 6.1   Statistics showing the status of autism across the world

No.	Country	Sample of population	Age group	Percentage	Ref.
1	Belgium	60 in 10,000	3–39 months	0.6	[14]
2	Canada	106 in 10,000	Under 18 years	1.06	[15]
3	China	23 in 10,000	Under 18 years	0.23	[16]
4	Estonia	60 in 10,000	Under 18 years	0.6	[17]
5	Denmark	69 in 10,000	Under 18 years	0.69	[18]
6	Finland	54 in 10,000	Under 8–26 months	0.54	[19]
7	Germany	38 out of 10,000	Under 18 years	0.38	[20]
8	Hong Kong	371 in 10,000	5–19 months	0.24	[21, 22]
9	Ireland	153 in 10,000	Under 18 years	1.53	[23]
10	Japan	181 in 10,000	Under 18 years	0.181	[24]
11	Netherland	48 in 10,000	Under 18 years	0.48	[25]
12	Norway	51 in 10,000	6–12 months	0.51	[26]
13	Poland	3 in 10,000	Under 18 years	0.34	[27]
14	Singapore	67 in 10,000	Under 18 years	0.67	[28]
15	South Korea	263 in 10,000	7–12 years	2.63	[29]
16	Switzerland	145 in 10,000	Under 18 years	0.8	[30]
17	Switzerland	145 in 10,000	Under 27–30 months	1.45–2.33	[31]
18	Taiwan	5 in 10,000	Under 18 years	0.5	[32]
19	Taiwan	5 in 10,000	26.4% of 0–24 months	0.05	[33]
20	USA	222 in 10,000	3–17 years	2.22	[34]
21	India	1 in 125; 1 in 85	3–6 years; 6–9 years	0.90 in rural areas; 0.60 in hilly areas; 1.01 in urban areas; 0.1 in tribal areas; 0.61 in coastal areas	[69, 70]

As of now there are no medicines available for treating autism, but there are some medicines available for the treatment of the side effects caused by ASD, such as wretchedness, seizures, sleep deprivation, and issue centering.

6.3.1 World Status on Autism

Presented here is the status of the population with ASD population in different parts of the world.

6.4 Artificial Intelligence and Machine Learning

Man-made reasoning is uncommonly wide in range. As per Moore [35, 49], who was a dean at the School of Computer Science at CMG University, "Man-made reasoning is the science and designing of causing PCs to carry on in manners that, as of not long ago, we thought required human insight."

Lipton, an academician and researcher, explains on Approximately Correct [50], the term "artificial intelligence," or AI. According to him, AI "is optimistic, a moving objective dependent on those abilities that people have however which machines don't." AI, likewise, incorporates an impressive proportion of innovative progresses that we know. Earlier works of AI include examples such as the chess playing Deep Blue, the AI that vanquished the 1997 world chess champion, Gary Kasparov, utilizing a tree search calculations technique [51] to assess a huge number of moves every step of the way [52–55].

Man-made intelligence, as we are aware, today is symbolized by human-AI collaboration contraptions—such as Siri by Google and Alexa by Amazon—and by the AI-fueled video expectation frameworks that power Netflix, Amazon, and YouTube. These innovative headways are becoming dynamically significant in our day-to-day lives. Truth be told, they are astute partners that upgrade our capacities as people and experts, making us increasingly profitable.

6.4.1 Machine Learning

AI (machine learning) is part of man-made consciousness, and it is characterized by computer professional and AI pioneer Mitchell as follows: "Artificial Intelligence is the investigation of PC calculations that permit PC projects to consequently improve through involvement" [56]. Machine learning is one of the manners in which we hope to accomplish AI. AI depends on working with small to large amounts of information by analyzing and contrasting the information with regular examples and investigating subtleties.

For example, if you furnish an AI model with a great deal of tunes that you appreciate, along with the related sounds (move capacity, instrumentality, beat, or classification), it will have the option to mechanize (depending on the directed AI model utilized) and create a recommender framework [57, 58] to propose other music later from Netflix, Spotify, and other organizations that (with a high level of likelihood) you'll appreciate [58, 59].

In a basic model, if you load an AI program with a huge amount of information of x-beam pictures alongside their portrayal (indications, things to consider, and so forth), it will have the ability to help (maybe automate) investigate the information of x-beam photos later on. The AI replica will take a gander at every single one of the photos in the differing informational collection and discover regular examples found in pictures that have been marked with

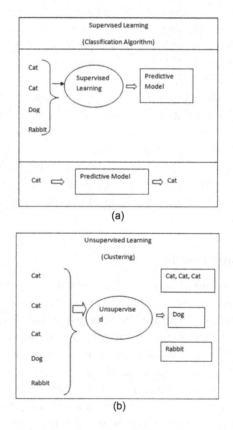

Figure 6.1   (a) Supervised learning and (b) unsupervised learning [60].

practically identical signs. Moreover, (expecting that we utilize a decent machine learning estimate for pictures) if a new set of data or images is loaded, features will be extracted from them and compared with trained examples to uncover how likely the images are dissected already.

When you load the model with new pictures, it will compare its parameters with the examples it has gathered before in order to disclose to you how likely are the pictures to contain any of the indications it has analyzed previously.

In supervised learning (Fig. 6.1), the algorithms build the relationship and calculate the dependencies connecting the target prediction output and the features fed at the input so that when a new data set is fed, it will predict the values using the previously learned relationships [61]. Solo learning, another kind of AI, is a group of AI calculations that are for the most part utilized in discovery and displays. These calculations don't have yield classifications or marks on the information (unlabeled data are used).

Reinforcement learning, another prevalent form of AI, utilizes perceptions assembled from the collaboration with its condition to take activities that would expand the reward or limit the hazard (Fig. 6.2). For this situation, the fortification learning calculation (called the operator) consistently gains from its condition utilizing

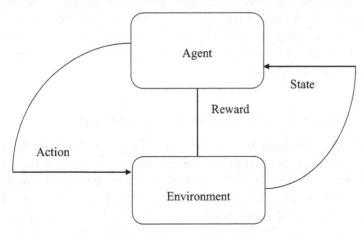

Figure 6.2   Reinforcement learning [52].

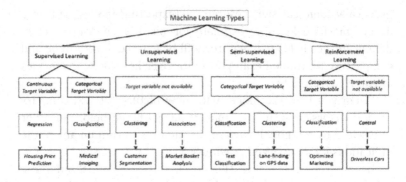

Figure 6.3   Types of machine learning [52].

cycle. An extraordinary case of support learning is a PC arriving at a super-human state and beating people at PC games [52].

Figure 6.3 shows the different machine learning algorithms and their classifications.

AI is entrancing, especially its subbranches, for example, profound learning and the different sorts of neural systems. Regardless, it is "enchantment" (computational learning theory), paying little mind to whether general society now and again has issues watching its interior activities. Indeed, while some will, in general, contrast profound learning and neural systems with the manner in which the human cerebrum works, there are significant contrasts between the two [62, 63]. Computational learning techniques are used in medical research, such as classification, and also in wound tissue characterization [71, 72].

6.4.2 Interchange of AI and Machine Learning

A gathering of analysts, including Simon and Newell [64], in 1956, shed light on the expression "man-made reasoning." From that point on, this industry has experienced numerous ups and downs. In the preceding decade, there was a ton of publicity around the business and numerous researchers agreed that AI matching the intelligence of humans was practically around the corner. But undelivered attestations upset people and prompted an "AI winter," a period where financing of and enthusiasm for the field died down significantly [65].

A while later, associations endeavored to isolate themselves from the term "artificial intelligence," which had turned out to be synonymous with unverified publicity, and used various entities to allude to the work they presented. For example, IBM portrayed a supercomputer called Deep Blue, which expressly stated that it didn't utilize man-made consciousness, when in reality, it did [66].

During this period, an assortment of different terms, for example, enormous information, prescient examination, and AI, began to gain popularity and prevalence. In 2012, AI, neural systems, and learning algorithms made extraordinary strides and began being used in a number of fields. Associations, all of a sudden, began to utilize the terms "AI" and "profound," figuring out how to promote their items.

Deep learning began to carry out errands difficult to accomplish with exemplary standard programming. Fields, for example, face and speech recognition, classification of images, and natural language processing (NLP), which were in a nascent stage, all of a sudden took incredible jumps [68]. In March 2019, three of the perceived profound learning pioneers won a Turing grant because of their commitments and achievements that have made deep neural systems a basic part of computational learning nowadays [67].

6.5 Autism, AI, and Machine Learning

From investigations it is clear that it is hard to identify autism in the beginning stages and control it. In writing, AI calculations are utilized for early, more precise prediction of ASD.

For some youngsters with a mental imbalance (ASD), perceiving and reacting to eye-to-eye connection, nonverbal communication, and manner of speaking is a noteworthy test. Improving these social aptitudes can take loads of work—putting a strain on parental figures with constrained time, assets, and cash for treatment.

Scientists have long realized that robots—and games with computerized criticism—can change the conduct of youngsters with a chemical imbalance, at least for the time being. Such associations have appeared to enable youngsters to make meaningful gestures, for example, looking, that they may have missed learning from their parental figures. In any case, converting these new skills into better individual-to-individual collaborations may require longer and increasingly escalated preparation, and few examinations have been huge enough—or long enough—to demonstrate critical, enduring enhancements.

6.5.1 JIBO-Human ROBO

Scassellati, a mechanical technology master and subjective researcher at Yale University, set up an investigation that gave youngsters a long-haul association with their bots, one they could impart to their families. His group gave 12 families a tablet PC stacked with social games and an adjusted adaptation of an economically sold robot called Jibo, which was customized to track the games and give criticism. "As a roboticist, that was one of the most alarming things on the planet. Leaving the robots there and trusting they would do the things we'd modified them to do," Scassellati says.

For 30 minutes every day, during the 30-day study, the youngsters sat beside their folks or other relatives and messed around while Jibo interfaced with them. The games depended on clinical treatment strategies for improving distinctive social abilities, including social and passionate understanding, taking another's viewpoint, and finishing errands in an arrangement. For instance, in a game called Rocket, played on a PC tablet, the tyke and the parental figure alternately structure a rocket by hauling parts around the screen. The screen is then reset to conceal this structure, and the primary player must disclose to the second player how to reproduce the plan.

All through the games, Jibo would display great social conduct by focusing its eyes on the kid or the guardian and orienting its body toward them. It would advise the youngster to do likewise. "The general purpose of the robot," Scassellati says, "is to make the cooperation between the youngster and the parent better. Every one of the aptitudes that they're learning, they're figuring out how to coordinate . . . to the parent, not to the robot." At a few points, the robot would call the youngster by name, requesting that the youngster draw in his or her guardian by taking a gander at the guardian. After every session, the framework's product changed the game's trouble depending on the tyke's presentation.

6.5.2 Autism Study Using ROBO

Parental figures, likewise, scored youngsters on how they carried on around other individuals (including themselves) and how connected with others they appeared during the games. Specialists taped the sessions, searching for changes in nonverbal communication and meaningful gestures among the kids. Finally, the analysts tried the kids with "joint consideration," which estimates how well somebody can call someone's attention to an item by staring at it, pointing to it, or verbally. That testing happened multiple times: 30 days before the test, on the primary day of the games, on the most recent day of the games, and again 30 days after the finish of the mediation.

Parental figures' reports demonstrated that the social abilities of every one of the 12 kids improved through the span of the examination, as reported by the specialists in *Science Robotics*. They were increasingly receptive to correspondence, started more discussions, and looked at others. The youngsters' normal joint consideration scores additionally improved between the first day and after a few days with the robot, expanding by 33% before dropping marginally 30 days after the investigation finished.

Scassellati says he isn't amazed by the drop in scores in the wake of playing with the robots. "Regardless of whether I put the best advisor that I have in your home, we wouldn't hope to see a perpetual change after only 30 days." So far, the consequences of the treatment of autism with robots are the same as the consequences of the treatment of autism with chatbots. As far as changes over the long run, he says, "We can't state now that the robot really creates long haul, enduring social change, yet what we do see is promising."

Despite the fact that the investigation could show changes in each youngster's conduct, what it couldn't quantify is the manner by which the bot treatment works with respect to different medicines, says Elizabeth Broadbent, who concentrates on human-robot collaborations in social insurance at the University of Auckland, New Zealand. This is because every kid kept getting his or her normal ASD treatment during the examination.

Scassellati says more work is expected to test the viability of the robots with bigger gatherings of youngsters over longer time frames. He concludes that the robots were never proposed to supplant customary medicines and treatment but to amplify the adequacy of these rare assets. "Most families can't stand to have a specialist with them consistently," says Scassellati, "Yet we can envision having a robot that could be with the family consistently, constantly, on interest, at whatever point they need it."

6.5.3 Autism Prediction Using ML Algorithms

Wall et al. made a video rater using a versatile Web-based interface for conducting a survey using 30 conduct highlights, including eye-to-eye connection and social grin, utilized by eight autonomous AI models for distinguishing autism disorder that show a precision greater than 94% in a cross-approval test and ensuing free approval. Three raters incognizant with regard to the conclusion autonomously estimated each one of the 30 highlights from the 8 models, with a middle time to fruition of 4 minutes. However, a few models (comprising rotating choice trees, support vector machine [SVM], calculated relapse (logistic regression [LR]), spiral part, and direct SVM) performed well. A scanty 5-include LR classifier (LR5) yielded the most elevated precision (area under the curve [AUC]: 92% [95% CI 88%–97%]) across all ages tried. The raters utilized a tentatively gathered free approval set of 66 recordings (33 ASD and 33 non-ASD) and 3 autonomous rater estimations to approve the result, accomplishing lower yet similar exactness (AUC: 89% [95% CI 81%–95%]). Finally, they applied LR to the 162-video-include grid to build an 8-highlight model, which accomplished 0.93 AUC (95% CI 0.90%–0.97%) on the held-out test set and 0.86 on the approval set of 66 recordings. Approval on kids with a current conclusion constrained the capacity to sum up the exhibition to undiscovered populaces [44].

6.5.4 Chatbot

A chatting bot, also called a chatbot, is a piece of programming that resembles a person. The bot will react to the client's content-based inquiries or sentences as a human. A useful bot can react to the contents in a way very similar to a person's reactions. The historical backdrop of improvement in chatbots is the length of the programming. While chatbots are presently available for various devices, from personal computers to smartphones (Android and Apple iPhone), their fundamental utilization is in diversion, while using the talking application based on the virtual help operator present at different sites [36]. There is decent space for further study and research in different areas for utilizing chatbots, and brain research is one such area.

At this time, there are countless counterfeit visit bots accessible. The primary goal of these talk bots is constrained to stimulation, customer backing, and promotion [37]. Some of the chatbots are additionally utilized for showing reason and better learning of understudies [38–40]. This ends up being a fascinating point. That is, we can utilize an explicitly planned chatbot to collaborate with the client, get more data with regard to his or her brain research, and analyze his or her mental issues if such issues exist. Mobile applications have been developed to help the public, hospitals, and doctors save time by collecting crucial patient information securely. Also the information can be seen without the patient having to carry the reports every time to the hospital, an advantage being that the patient need not fear losing the files. Moreover, the information can be accessed anywhere, anytime with more confidentially, without risking the privacy of the patient [73].

On other side, autism can be identified and treated using AI. In Ref. [41], facial automation for conveying emotions, which is also called "FACE," connects with kinesics (or nonverbal correspondence passed by body part developments, or outward appearances, etc.) and considers the polemics space, which characterizes and tests a restorative convention for mental imbalance so as to improve social and emotive capacities in individuals. Individuals with a chemical imbalance concentrate on single subtleties, yet the association with a robot may enable a medically introverted subject to focus on himself or herself on the predetermined quantity of robot. Authors in Ref. [42] focused on improving the benchmarks of autism diagnosis by utilizing special social robots to give quantitative target estimations of reactions given by the person. Intelligent social robots that are also interactive are used to elicit a specific reaction. This helps create a standard procedure to record and calculate the reactions. It finishes up with an exchange on remedial and indicative potential outcomes for this work and conjectures on how the utilization of social robots in autism research may prompt a more noteworthy comprehension of this turmoil. A number of chatbots are now being utilized in psychiatry [43].

At the University of Barcelona, a chatbot has been created and trained for diagnosing generalized anxiety disorder. It also helps improve the skills of the students of psychology. An embodied conversational agent is one of the developed chatbots inspired from the architecture developed in the University of Barcelona that can express feelings and character attributes using composed writings. It utilizes and displays the feelings and characteristics invested in it, dependent on AI Markup Language.

There are a few impediments that existing frameworks face. The master frameworks presented in Refs. [37, 45] have mental utilization with impediment of bouncing client to react in some given alternatives. These master frameworks do not give clients the opportunity to state anything they desire to state, which can be progressively useful in drawing a genuine image of a client's mental issues. A client will be unable to cooperate with these master frameworks in a loosened-up condition.

A chatbot helps in the following conditions by gauging the reason for the way a client is influenced before a genuine specialist is included.

(1) A table is created in a Microsoft Excel sheet that can be used as a database at any chatbot stage or utilizing python, a chatbot can be made that is tweaked to our needs.
(2) The chatbot is a conversational bot that associates with the patient effectively.
(3) Images associated with the ASD individual are used.
(4) Conversations are further used by the doctor.

6.5.4.1 Algorithm to create a simple chatbot

Here is an algorithm that helps create a simple chatbot:

(1) Start.
(2) Arbitrarily choose the message (user message or chatbot's).
(3) Hang around for the client's information.
(4) Respond with a solution.
(5) Divide the given message into sentences.
(6) Divide each sentence into tokens.
(7) Acquire equivalent words for each token.
(8) Utilize equivalent words for producing new sentences.
(9) Explore the inquiries in the cerebrum.
(10) If a query is discovered at this point, produce a reply and put it on screen.
(11) If no query is discovered, pose the bot's individual inquiry. Else return to Step 2.
(12) End.

6.5.4.2 Process to create a chatbot

Here are the steps for creating a chatbot:

(1) NLP: This is used to process the response to the query posed by the user in the diagnosis process [46].
(2) Tokenization: The sentence is split into tokens, and these tokens are put aside for correlation.
(3) Extraction of keyword: Every token is verified and contrasted with the words that can clarify force of any manifestation or watchwords.
(4) Similarity coordination of a sentence: After every token correlation, the entire sentence's closeness is coordinated for affirming whether the watchwords are utilized in a similar setting in order to clarify the manifestation power.
(5) Understanding the importance of keywords: After comparability coordination, the catchphrases are mapped to the seriousness of the side effect; here four degrees of seriousness are utilized: ordinary, minor, normal, and serious.

6.5.4.3 Framework of the chatbot

Figure 6.4 shows the framework of a chatbot.

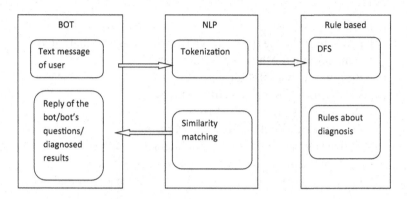

Figure 6.4   Framework of the chatbot.

Here is a sample database of questions to be framed:

- What is your name?
- How are you doing?
- Had your breakfast?
- Oh! It is already time for breakfast.
- Taken the medicines in the morning?
- You have earned a star for this.

The framework comprises various modules, which can be partitioned into three center modules:

- Generate answers to the client's questionnaires.
- Process client information and navigate the choice tree until you reach the leaf.
- The decision tree itself.

Flow of system: Here is how the framework stream will be:

(1) The user will enter an instant message into the visit window.
(2) The message will be divided into sentences.
(3) Independently, each sentence will be further split into expressions.

(4) Synonyms of expressions will be taken from information base; new sentences will be produced utilizing equivalent words (maximum 20–30 sentences); search will be made for inquiries in mind.

Decision trees are quick, effectively executed, and easy to get. They are one of the least complex basic leadership strategies accessible. The augmentations to essential model can make them amazingly advanced and ground-breaking. They have the upside of being secluded and simple to make. They can be seen being utilized for everything, from activity to complex vital and strategic AI [45]. They can likewise be scholarly, and that learning is generally quick when contrasted with methodologies, for example, neural systems or hereditary calculations. They can be utilized in chatbots to improve their presentation [47]. Given a lot of learning, we have to create a related activity from a lot of potential activities. The mapping among information pieces and yield might be mind-boggling. A similar activity will be utilized for sets of a wide range of information. However, any little change in one piece of information may make an activity seem reasonable or dumb. For instance, on the off-chance that we ask the client whether he or she knows cricket, the client's answer might be in the affirmative or negative. If it is in the affirmative, it bodes well to ask whether the client likes bowling or batting. However, on the off-chance that the answer is negative, posing a similar inquiry will be useless and moronic.

In psychological chatbots, as portrayed in the System Design parcel, the framework needs to settle on a ruling against every single contribution from the client. So basic leadership is the center segment of our AI model. The thought is that the client will enter the content and the framework will process it. Now the framework, likewise, needs to choose what to ask next in the discussion. This leadership is significant since it is concerned with legitimate and consistent working of the above framework; else it will look peculiar or maybe awful. We use decision trees for our basic leadership prerequisites on account of their straightforwardness and power. A conceivable structure of a choice tree can be found in Fig. 6.5.

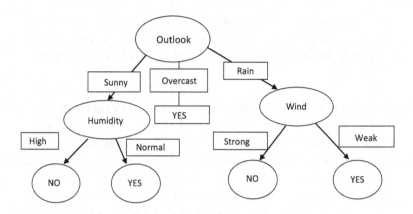

Figure 6.5 An example of a decision tree.

One of the generated questions will match one of the answers at random. Otherwise, if no question matches, the topic will be changed.

6.5.5 Role of AI Chatbots

AI chatbots have become one way of increasing business by providing customer service. These chatbots can respond to customer queries in real time, at any time, in a fast, dynamic, and effective way. They are also cheap and easy to maintain. The major point is that it is easy to deploy a chatbot and takes only a few minutes to integrate. There are also a lot of advantages in using AI chatbots for service [48].

The effectiveness of these chatbots is in that they can predict queries and prepare suitable responses. A teacher at Georgia Tech University made a chatbot to use as an educating right hand. It addressed understudies' inquiries for a figuring class [49]. In the quarter of a year the chatbot could respond to inquiries with 97% precision.

Autistic persons require trained personnel to take care of them. The parents are also trained on how to interact with them. But the reality is that it is difficult for middle-income-group families to have trained personnel. Here, it must be mentioned that autism is present across the world. One of the main problems identified in countries like India and other developing countries is that parents of such children do not come out readily to speak to doctors because of the possible social stigma. The child has to spend most of his or her time within the confines of the house, without any interaction with the outside world. As such problems are handled by professionals only, often, parents of such children are also unable to bear the cost of medications and doctor fee.

Here, we list some solutions to address the problems keeping in mind all sections of people in the world:

- Creating a simple chatbot to diagnose the disease and save the data
- Creating a simple dashboard where the doctor (helper) can send text messages to the patient directly and can interact with him or her
- Creating a dashboard with the chatbot for the doctor to interact with the patient on the Web page
- Creating a 3D chatbot with AI features using machine learning algorithms to make it learn and interact with the patient so that the doctor need not spend the time to chat and also can check afterward the chat details

Research is on to create interactive robots so that patients can play with such robots. These robots also change colors with the moods of the patients. But poor people cannot afford such robots. Also, this type of technology may not reach the third world and other developing countries at a low cost. And even if such a robot is sponsored by someone for a poor family, its maintenance can be an issue.

6.5.6 Chatbot Model

There are four stages to creating the perfect 3D chatbot model:

(1) Creating a chatbot for diagnosis
(2) Creating a simple text chat
(3) Creating a dashboard
(4) Creating a 3D chatbot model

6.5.6.1 Creating a chatbot for diagnosis

The first stage is creating a chatbot for diagnosis. Figure 6.6 shows the model of a chatbot that will help diagnose ASD.

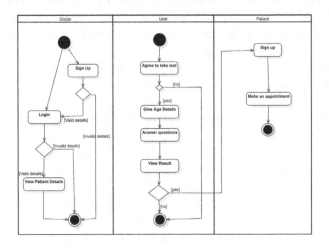

Figure 6.6   Flow of the developed model.

A chatbot model of this type will show the following possible screens when the user interacts with it.

(1) This is the welcome screen—the dashboard (Fig. 6.7).

Figure 6.7   Developed welcome screen.

(2) This screen will show the patients assigned to the doctor. It can be populated through the database (Fig. 6.8).

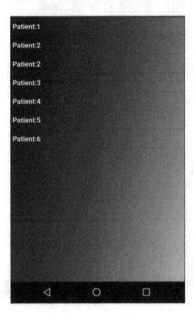

Figure 6.8   Figure showing the list of patients registered.

(3) This is the user screen—the test portal where the user will enter the details of the patient (Fig. 6.9).
(4) As the user answers each query, more questions will appear (Fig. 6.10).
(5) This is the result screen when the ASD symptoms are high (Fig. 6.11).

After the test, the doctor can verify the data, perform counseling, and with the help of the combined information, confirm ASD and start the patient on medication.

6.5.6.2 Creating a simple text chat

As we discussed earlier, in the second stage, we prepare a chatbot that will have direct contact with the child or the parent so that he or she need not come daily to the doctor. This will help, especially

Figure 6.9 Figure showing the details to be entered.

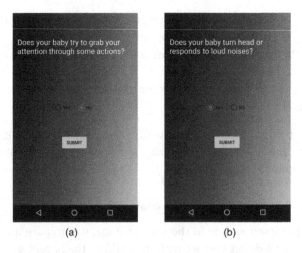

(a)   (b)

Figure 6.10 Figures showing sample questions parents have to answer and the responses recorded.

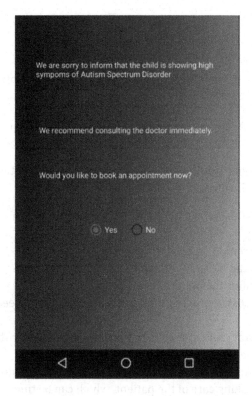

Figure 6.11   Figure showing the result after the test is taken.

in developing countries like India, where parents do not like to talk about their children's condition and do not let them go out and many people in rural areas do not have access to specialists, who are usually found in cities only. These types of tools can bridge the gap between them and the burden on the parents also decreases economically.

As the doctor has access to continuous evaluation, he or she can make decisions on medication also. The point is that the doctor is also busy and cannot simply chat with the patient continuously, even when required. This model can chat automatically with the patient using machine learning algorithms and also save the conversations with the patient in the database. The doctor now can access this data and can analyze the condition of the patient.

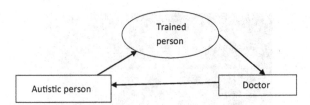

Figure 6.12 Traditional interaction between an autistic person and a doctor.

Here is what traditional interaction between an autistic person and a doctor involves (Fig. 6.12):

- An autistic patient needs to be taken proper care of by the parents.
- Parents require special training for it.
- Parents can be busy with work and may have to hire a caretaker to look after the patient.
- It is a costly process, and the caretaker also needs proper training.
- The doctor can get feedback from the caretaker or the parents for further evaluation of the disorder.
- The doctor can make decisions based only on feedback of the person taking care of the patient, which can be time consuming. Also, sometimes, some minute but important things can be missed by this person and not conveyed to the doctor, which can affect the medication process of the patient.

Here is what a chatbot interaction between an autistic person and a doctor involves (Fig. 6.13):

- Instead of a human (caretaker), the doctor can interact with the patient at any time anywhere through the chatbot.

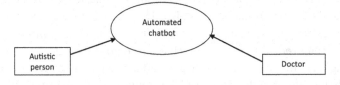

Figure 6.13 Chatbot interaction between an autistic person and a doctor.

- Since it is automated, it can chat with the patient as it has trained by using machine learning algorithms.
- Each moment is recorded. This is helpful for the doctor as the chatbot notes each log and every chat and, thereby, also records the moods and emotions of the patient. The doctor can easily refer to the log and need not depend on anyone for information on the patient.
- On the basis of these recordings, the doctor can make proper decisions regarding the medication of the person.
- Errors are minimized in this interaction, compared with interaction involving a human caretaker.
- This also reduces the cost for the parents and is more effective for parents and doctors.

6.5.6.3 Creating a dashboard and a 3D chatbot model

The third stage is to maintain a dashboard on the system so that it becomes convenient for the doctor. The fourth stage is creating a 3D model for chatting for better interaction, one that is affordable even by normal persons.

6.6 Conclusion

In this chapter, ASD and its ill effects have been studied. No medicines are available for treating autism, but there are some symptoms of autism that can be cured with the help of medicines. AI and machine learning have shown great progress and can also be used to predict autism at an early stage. AI chatbots can be designed to interact with children and their doctors. They can also chat with the child when the doctor is not available. Interactive robots with emotions have been designed to interact with people with ASD. These not only help to reduce the burden on the parents or caretakers but also help the doctor to treat an autistic person as all emotions can be recorded and proper reading can be done and decision made by the doctor.

6.7 Future Scope

In developing countries, parents are afraid to expose their autistic children to the society or discuss the problem with or even meet the doctors regarding the problems faced by their children. Hence it is difficult in developing countries to screen for the disease at an early stage. Nowadays, a lot of literature on ASD is available on the Internet and applications for smartphones and the Internet can help to screen for the disease at an early stage. In the future, screening for autism could also be done using videos of the patient rather than the patient taking the test, which will avoid the issue of fluke answers given at the time of answering the questions. The videos of the children in different situations could be uploaded, and using these videos, doctors could screen for ASD. There is also a need to extend the services of these applications so that they help control disorders that occur due to autism. The enhanced system should warn the parents of the consequences and be able to detect the disorders asking the questions related to those disorders, making it easy for doctors to screen the patients.

References

1. American Psychiatric Association (2013). *Diagnostic and Statistical Manual of Mental Disorders*, 5th ed. Washington, DC.
2. Jang, J., Dixon, D. R., Tarbox, J., Granpeesheh, D. (2011). Symptom severity and challenging behavior in children with ASD. *Res. Autism Spectr. Disord.*, **5**:1028–1032, https:// doi.org/10.1016/j.rasd.2010.11.008.
3. Matson, J. L., Wilkins, J., Macken, J. (2009). The relationship of challenging behaviors to severity and symptoms of autism spectrum disorders. *J. Mental Health Res. Intellect. Disabl.*, **2**:29–44, https:// doi.org/10.1080/19315860802611415.
4. Osofsky, J. D., Fitzgerald, H. E. (2000). *WAIMH Handbook of Infant Mental Health*, Vol. 1 (Perspectives on Infant Mental Health). Wiley, Chichester, UK.
5. https://www.who.int/news-room/fact-sheets/detail/autism-spectrum-disorders.

6. https://www.autismspeaks.org/technology-and-autism.

7. Baio, J., Wiggins, L., Christensen, D. L., et al. (2018). Prevalence of autism spectrum disorder among children aged 8 years — autism and developmental disabilities monitoring network, 11 sites, United States, 2014. *MMWR Surveill. Summ.*, **67**(6):1–23, http://dx. doi.org/10.15585/mmwr.ss6706a1external icon.

8. https://www.helpguide.org/articles/autism-learning-disabilities/does-my-child-have-autism.htm.

9. https://www.ninds.nih.gov/Disorders/Patient-Caregiver-Education/Fact-Sheets/Autism-Spectrum-Disorder-Fact-Sheet.

10. Chaste, P., Leboyer, M. (2012). Autism risk factors: genes, environment, and gene-environment interactions. *Dialogues Clin. Neurosci.*, **14**(3):281–292.

11. http://www.cdc.gov/ncbddd/autism/screening.html.

12. Leitner, Y. (2014). The co-occurrence of autism and attention deficit hyperactivity disorder in children - what do we know? *Front. Hum. Neurosci.*, **8**:268, doi:10.3389/fnhum.2014.00268.

13. Leigh, J. P., Du, J. (2015). Brief report: forecasting the economic burden of autism in 2015 and 2025 in the United States. *J. Autism Dev. Disord.*, **45**(12):4135–4139, doi:10.1007/s10803-015-2521-7.

14. Dereu, M., Warreyn, P., Raymaekers, R., Meirsschaut, M., Pattyn, G., Schietecatte, I., Roeyers, H. (2010). Screening for autism spectrum disorders in Flemish day-care centres with the checklist for early signs of developmental disorders. *J. Autism Dev. Disord.*, **40**(10):1247–1258, doi:10.1007/s10803-010-0984-0.

15. https://www.autismontario.com/client/aso/ao.nsf/web/Info+about+ASD?OpenDocument.

16. Wan, Y., Hu, Q., Li, T., Jiang, L., Du, Y., Feng, L., Wong, J. C.-M., Li, C. (2013). Prevalence of autism spectrum disorders among children in China: a systematic review. *Shanghai Arch. Psychiatry*, **25**(2):70–80, doi:10.3969/j.ISSN.1002-0829.2013.02.003.

17. https://intra.tai.ee/images/eventlist/events/13_HPH13_Childrens_mental_health_Kleinberg.pdf.

18. Parner, E. T., Thorsen, P., Dixon, G., de Klerk, N., Leonard, H., Nassar, N., Bourke, J., Bower, C., Glasson, E. J. (2011). A comparison of autism prevalence trends in Denmark and Western Australia. *J. Autism Dev. Disord.*, **41**(12):1601–1608, doi:10.1007/s10803-011-1186-0.

19. Hinkka-Yli-Salomäki, S., Banerjee, P. N., Gissler, M., Lampi, K. M., Vanhala, R., Brown, A. S., Sourander, A. (2014). The incidence of diagnosed autism spectrum disorders in Finland. *Nord. J. Psychiatry*, **68**(7):472–480, doi:10.3109/08039488.2013.861017.

20. Bachmann, C. J., Gerste, B., Hoffmann, F. (2018). Diagnoses of autism spectrum disorders in Germany: time trends in administrative prevalence and diagnostic stability. *Autism*, **22**(3):283–290, https://doi.org/10.1177/1362361316673977.

21. http://www.scmp.com/news/hong-kong/education-community/article/2081935/hong-kong-neglecting-needs-its-autistic-pupils.

22. https://en.wikipedia.org/wiki/Demographics_of_Hong_Kong.

23. http://www.thejournal.ie/autism-irish-schools-2879550-Jul2016/.

24. Kawamura, Y., Takahashi, O., Ishii, T. (2008). Reevaluating the incidence of pervasive developmental disorders: impact of elevated rates of detection through implementation of an integrated system of screening in Toyota, Japan. *Psychiatry Clin. Neurosci.*, **62**(2):152–159, doi:10.1111/j.1440-1819.2008.01748.x.

25. van der Ven, E., Termorshuizen, F., Laan, W., Breetvelt, E. J., van Os, J., Selten, J. P. (2013). An incidence study of diagnosed autism-spectrum disorders among immigrants to the Netherlands. *Acta Psychiatr. Scand.*, **128**(1):54–60, doi:10.1111/acps.12054.

26. Isaksen, J., Diseth, T. H., Schjølberg, S., Skjeldal, O. H. (2012). Observed prevalence of autism spectrum disorders in two Norwegian counties. *Eur. J. Paediatr. Neurol.*, **16**(6):592–598, doi:10.1016/j.ejpn.2012.01.014.

27. Piskorz-Ogórek, K., Ogórek, S., Cieślińska, A., Kostyra, E. (2015). Autism in Poland in comparison to other countries. *Pol. Ann. Med.*, **22**(1):35–40, https://doi.org/10.1016/j.poamed.2015.03.010.

28. http://www.straitstimes.com/singapore/health/1-in-150-children-in-singapore-has-autism.

29. Kim, Y. S., Leventhal, B. L., Koh, Y. J., Fombonne, E., Laska, E., Lim, E. C., Cheon, K. A., Kim, S. J., Kim, Y. K., Lee, H., Song, D. H., Grinker, R. R. (2011). Prevalence of autism spectrum disorders in a total population sample. *Am. J. Psychiatry*, **168**(9):904–912, doi:10.1176/appi.ajp.2011.10101532.

30. https://www.swissinfo.ch/eng/report-calls-for-end-to-neglect-of-autism/4006988.

31. https://en.wikipedia.org/wiki/Demographics_of_Switzerland.

32. Hsu, S.-W., Chiang, P.-H., Lin, L.-P., Lin, J.-D. (2012). Disparity in autism spectrum disorder prevalence among Taiwan National Health Insurance enrollees: age, gender and urbanization effects. *Res. Autism Spectr. Disord.*, **6**(2):836–841.

33. http://www.indexmundi.com/taiwan/demographics_profile.html.

34. https://www.cdc.gov/nchs/data/nhsr/nhsr087.pdf.

35. https://blogs.wsj.com/cio/2018/07/27/what-machine-learning-can-and-cannot-do/.

36. https://www.forbes.com/sites/peterhigh/2017/10/30/carnegie-mellon-dean-of-computer-science-on-the-future-of-ai/#164487 aa2197.

37. Nunes, L. C., Pinheiro, P. R., Pequeno, T. C. (2009). An expert system applied to the diagnosis of psychological disorders. In *2009 IEEE International Conference on Intelligent Computing and Intelligent Systems*, Shanghai, pp. 363–367.

38. Sun, B., Kang, W., Zhang, R., Fang, Z., Xu, X. (2010). PsyCare: a novel framework for online psychological counseling. In *IEEE 15th International Confrence on Pervasive Computing and Applications (ICPCA)*, pp. 56–61.

39. Huang, Y.-T., Yang, J.-C., Wu, Y.-C. (2008). The development and evaluation of English dialogue companion system. In *8th IEEE International Conference on Advanced Learning Technologies, ICALT 2008*, Santander, Cantabria, Spain.

40. Niranjan, M., Saipreethy, M. S., Gireesh Kumar, T. (2012). An intelligent question answering conversational agent using naïve bayesian classifier. In *IEEE International Conference on Technology Enhanced Education (ICTEE)*, pp. 1–5.

41. Pioggia, G., Sica, M. L., Ferro, M., Jgliozzi, R., Muratori, F., Ahluwalia, A., De Rossi, D. (2007). Human-robot interaction in autism: FACE, an android-based social therapy. In *RO-MAN 2007 - The 16th IEEE International Symposium on Robot and Human Interactive Communication*, Jeju, pp. 605–612.

42. Scassellati, B. (2005). Quantitative metrics of social response for autism diagnosis. In *ROMAN 2005. IEEE International Workshop on Robot and Human Interactive Communication, 2005*, Nashville, TN, USA, pp. 585–590.

43. Morales-Rodríguez, M. L., González, B. J. J., Florencia Juárez, R., Fraire Huacuja, H. J., Martínez Flores, J. A. (2010). Emotional conversational

agents in clinical psychology and psychiatry. In *Advances in Artificial Intelligence*, MICAI 2010. Lecture Notes in Computer Science, Vol. 6437, Springer, Berlin, Heidelberg, pp. 458–466.

44. Tariq, Q., Daniels, J., Schwartz, J. N., Washington, P., Kalantarian, H., Wall, D. P. (2018). Mobile detection of autism through machinelearning on home video: a development and prospective validation study. *PLoS Med.*, **15**(11):e1002705, https://doi.org/10.1371/journal. pmed.1002705.

45. Agarwal, S., Agarwal, P. (2005). A fuzzy logic approach to search results' personalization by tracking user's Web navigation pattern and psychology. In *Proceedings of the 17th IEEE International Conference on Tools with Artificial Intelligence (ICTAI'05)*, Hong Kong, pp. 318–325.

46. Wiak, S., Kosiorowsk, P. (2010). The use of psycholinguistics rules in case of creating an intelligent chatterbot. In *Artifical Intelligence and Soft Computing*, ICAISC 2010. Lecture Notes in Computer Science, Vol. 6114, Springer, Berlin, Heidelberg, pp. 689–697.

47. Casagrande, E., Woldeamlak, S., Woon, W. L., Zeineldin, H. H., Svetinovic, D. (2014). NLP-KAOS for systems goal elicitation: smart metering system case study. *IEEE Trans. Software Eng.*, **40**:941–956.

48. https://www.livechatinc.com/blog/chatbots-improve-customer-service/.

49. Moore, A. Artificial Intelligence and Machine Learning Department at Carnegie Mellon University, https://www.youtube.com/watch?v=HH-FPH0vpVE.

50. Lipton, Z. C. http://approximatelycorrect.com/2018/06/05/ai-ml-ai-swirling-nomenclature-slurried-thought/.

51. Chijindu, Engr. V. C. (2012). Search in artificial intelligence problem solving. *Afr. J. Comput., ICT*, **5**(5).

52. Fumo, D. retrieved on Jun 15, 2017 from https://towards datascience.com/types-of-machine-learning-algorithms-you-should-know-953a08248861.

53. Campbell, M. S., Hoane, Jr., A. J., Hsu, F.-h. (1999). Search control methods in deep blue. From *AAAI Technical Report SS-99-07* (www.aaai.org), pp. 19–23.

54. Metz, C. retrieved on October 22, 2017 https://www.nytimes.com/2017/10/22/technology/artificial-intelligence-experts-salaries. html.

55. Chris, Neils and Percy retrieved from Cade Metz http://stanford. edu/~cpiech/cs221/apps/deepBlue.html 2013.

56. Mitchell, T. (1997). *Machine Learning*. McGraw Hill, ISBN 0070428077.

57. Ricci, F., Rokach, L., Shapira, B. (2011). Introduction to recommender systems handbook. In *Recommender Systems Handbook*, Springer, pp. 1–35.

58. Gupta, P., Goel, A., Lin, J., Sharma, A., Wang, D., Zadeh, R. B. (2013). WTF: the who-to-follow system at Twitter. In *Proceedings of the 22nd International Conference on World Wide Web*, pp. 505–514.

59. Le, J. retrieved on July 11, 2018 https://towardsdatascience.com/spotifys-this-is-playlists-the-ultimate-song-analysis-for-50-mainstream-artists-c569e41f8118.

60. https://blog.westerndigital.com/machine-learning-pipeline-object-storage/.

61. Stanford, S., Iriondo, R. retrieved May 15, 2019 https://medium.com/towards-artificial-intelligence/the-50-best-public-datasets-for-machine-learning-d80e9f030279.

62. https://openai.com/five/.

63. Savage, N. (2019). How AI and neuroscience drive each other forwards. *Nature*, **571**:S15–S17, doi:10.1038/d41586-019-02212-4.

64. Simon, H. A. (1996). *The Sciences of the Artificial*. MIT Press Cambridge, MA, ISBN:0-262-69191-4USA.

65. Lim, M. Retrieved on September 5, 2018 https://www.actuaries.digital/2018/09/05/history-of-ai-winters/.

66. Campbell, M., Hoane, Jr., A. J., Hsu, F.-h. (2002). *Deep blue. Artif. Intell.*, **134**(1–2):57–83.

67. https://amturing.acm.org/byyear.cfm.

68. Minar, M. R., Naher, J. Recent advances in deep learning: an overview. https://arxiv.org/pdf/1807.08169.pdf.

69. Rudra, A., Belmonte, M. K., Soni, P. K., Banerjee, S., Mukerji, S., et al. (2017). Prevalence of autism spectrum disorder and autistic symptoms in a school-base cohort of children in Kolkata, *India. Autism Res.*, **10**:1597–1605.

70. Raina, S. K., Chander, V., Bhardwaj, A. K., Kumar, D., Sharma, S., et al. (2017). Prevalence of autism spectrum disorder among rural, urban, and tribal children (1-10 years of age). *J. Neurosci. Rural Pract.*, **8**:368–374.

71. Chakraborty, C. (2019). computational approach for chronic wound tissue characterization. *Inf. Med. Unlocked*, **17**:100162, https://doi.org/10.1016/j.imu.2019.100162.

72. Enireddy, V., Kumar, R. K. (2015). Improved cuckoo search with particle swarm optimization for classification of compressed medical images. *Sadhana*, **40**:2271–2285.

73. Chinmay, C. (2019). Mobile health (m-health) for tele-wound monitoring. In *Mobile Health Applications for Quality Healthcare Delivery*, IGI Global, Chapter 5, pp. 98–116, doi:10.4018/978-1-5225-8021-8.ch005.

Chapter 7

Emergence of Artificial Intelligence and Its Legal Impact

Amol Deo Chavhan

Manikchand Pahade Law College, Samarth Nagar, Aurangabad, Maharashtra 431001, India
amolchavhan13@gmail.com

7.1 Introduction

Through various marvelous inventions, science and technology reduce the pain and suffering of humans. Through computers and smart devices, human labor efforts are continuously reducing and may soon come to an end. People are smarter and faster owing to the Internet and other electronic devices. These devices and instruments work smartly due to their programing and design. Almost all smart devices are monitored and work through Artificial Intelligence (AI). In other words, AI reduces physical efforts of human beings. It reduces the pain of hard work by human beings. At the same time, people are more dependable physically and mentally. It is quite possible that in the near future, human intelligence will reduce as they come to depend more and more on AI.

Artificial Intelligence and the Fourth Industrial Revolution
Edited by Utpal Chakraborty, Amit Banerjee, Jayanta Kumar Saha, Niloy Sarkar, and Chinmay Chakraborty
Copyright © 2022 Jenny Stanford Publishing Pte. Ltd.
ISBN 978-981-4800-79-2 (Paperback), 978-1-003-15974-2 (eBook)
www.jennystanford.com

AI is not a new concept. It was developed and placed before people due to rapid development in science and technology. Earlier, people considered science and technology fiction, something the orthodox society did not believe in. Various inventions were rejected—rather not sanctioned by the so-called developed society. Now, AI is more science and less fiction.[1] With notions regarding AI changing across the world, human labor is being replaced by computers and other smart devices that work more smartly than humans. There is no possibility of errors or mistakes. AI manages the working of machines. In other words, AI is the brain of machines, using which machines work. It is an umbrella term that refers to information systems inspired by biological systems and encompasses multiple technologies, including machine learning (ML), deep learning, computer vision, natural language processing (NLP), machine reasoning, and strong AI.[2]

AI inventions are beginning to support humans in every sector, replacing the physical efforts of humans. AI is also replacing the mental activities of people. One can perform mental activities such as evaluation, mental state, and decision making using machines with AI. Sometimes, AI can also be used for criminal activities. As per the law, a person has criminal liability only if the act is committed with *mens rea* and *actus reus*. If the act is done through AI, the main hurdle before the court is to identify and punish the real culprit. Secondly, the "right to privacy" of a person may be hampered due to the use of AI.

If the computer, the mobile, and other electronic devices are held responsible for any criminal activity, the main hurdle is these devices are not considered to be within the ambit of what is defined as a "person." Offences and wrongs may be committed through these devices, and what may be proved is that the individual is not involved in wrongful activities. If the activities are not control

---

[1] Nishith Desai Associates. The future is here: artificial intelligence and robotics, May 2018, available at www.nishithdesai.com.

[2] PR Newswire. Artificial intelligence market forecasts, available at http://www.prnewswire.com/news-releases/artificial-intelligence-market-forecasts-300359550.html. cited in Nishith Desai Associates, The future is here: artificial intelligence and robotics, May 2018, available at www.nishithdesai.com.

AI will hamper education, agriculture, economy, health care system more precisely the fundamental right i.e. Right to Privacy. All the above-mentioned sectors are coming to gradually depend on AI. If anything goes wrong in any of these sectors, the wrongdoer has the liberty to cast the responsibility on the computer and other electronic devices used. Now, as per all our criminal and civil laws, only human beings can be held responsible for any crime. But AI does not come under the purview of a "human being," and if we consider it a corporate entity, then the question will arise as to "whom" to punish. In other words, the existing laws are not sufficient to punish the machines that commit a crime and may, in fact, end up violating the fundamental rights of individuals in such instances. For example, if we consider the person who operates the AI machine responsible for punishment/compensation, it may be great injustice to him or her because AI is operated by a "program," which may have been designed by somebody else.

The thing is that all machines are operated by humans. The *mens rea* can be implemented through machines, which will not implicate any person. The present study will evaluate different views and interpretations regarding whether human laws can be imposed on AI objects and how criminal liability and constitutional liability will be fixed on the particular person/machine. This study endorses the application of the interpretation of criminal law and constitutional rights by the court to detect whether the AI object is capable of showing *mens rea*. Before discussing the liability of machines, it is necessary to discuss and elaborate on the concept of AI.

7.2 What Is Artificial Intelligence?

AI plays a vital role in civilization today. It reduces the burden on humans through mechanical advancement. AI covers a wide range of sciences and technologies. There is no specific definition of AI. It developed along with the development of computers and smart devices. This means that the old concept of AI was overtaken by new and modern computer applications. In 1955,

McCarthy[3] attempted to define AI. As per his contention, It is the science and engineering of making intelligent machines, especially intelligent computer programs."[4] In terms of computer programs, this means that AI is a type of machine that works on some program. This also means that AI includes all kinds of machines and devices that work as per specific programs or are operated through computer software. Luger and Stubblefield defined AI "as the branch of computer science that is concerned with the automation of intelligent behavior."[5] AI refers to the intelligent behavior of computers and other smart devices.

AI is based on computer programs that lead the machine to work as per the program installed on it. In other words, one can say that AI is a program fixed by the programmer to operate a machine. It is a smarter way to handle the machine and reduce the pain and the hard time the machine operator would otherwise face. Making or introducing AI is like having a child, where the child will obey the parent and act as per the direction of the parent. AI is the systematic creation of a program that will be followed by the machine when performing its task. The machine is a child, and AI is the parent who inculcates good habits in it and nourishes these habits.

It is hard to define AI in a water-tight manner. The term "intelligence" has different meanings and notions, which can change as per the changing notions. No one can say with any authority that this thing or that thing is intelligence. It may vary from situation to situation and place to place. What is considered intelligence today may not be considered so in the future. The notion of intelligence changes in the fraction of a moment. AI is strict engineering invention and scientific approach. And it is for this reason that the answer to what exactly AI is is vague. One can say that AI is a powerful tool for exploring a scientific question or problem. And for this reason the definitions given by various jurists fall short and are ambiguous. AI is a new and emerging concept, and

---

[3]He was an American computer scientist known as the founder of artificial the intelligence discipline.
[4]Rajaraman, V. (2014). John McCarthy: father of artificial intelligence. *Resonance*, **19**:198–207.
[5]Luger, G. F. (2009). *Artificial Intelligence: Structures and Strategies for Complex Problem Solving*, 6th ed., Addison-Wesley.

with developments in science and technology, the structure, design, manner of work, and instruments of AI change daily. Additionally, the methods and working manner of AI change and for this reason, it is difficult to define AI in a straightforward manner.

7.3 Why Artificial Intelligence Is Necessary for Study?

Studying AI is necessary to find out about the mechanism of machines and smart devices. People think that machines work to reduce the pain of hard word for people, but in fact this task is performed by AI. Machines are useless if the software or program fails to work. AI makes it possible for machines to work. In other words, AI plays a crucial role in how machines function. Today, AI helps people to work smartly in every sector. At the same time, it places some legal and practical problems before humans. It is necessary to find out what these problems are and solutions for the same, and for this reason, the study of AI is necessary.

AI covers a vast range of sectors and all kinds of smart devices because it thinks and works like a human. It has changed human life in all aspects. With the use of these devices, human effort and labor have reduced. In other words, machines produce slaves to perform various tasks and work. Through AI, various activities are performed with maximum possible speed and accuracy. AI performs errorless and faster work.

AI is part of, for example, medical, industrial, banking, and aeronautic engineering sectors. The use of AI in these sectors has proved effective and cost efficient. Eventually, AI may help maximize production with minimum investment and efforts in these sectors. So far as the medical sector is concerned, with the help of AI, doctors and hospitals are able to make accurate diagnosis about diseases and give the right direction and medication to patients. Not only this, through AI, medical negligence cases have also reduced. In the banking sector, AI has helped to reduce queues and provide cheap and efficient banking services to customers. Additionally, in the big industries, AI has increased the productivity of various companies while helping curtail expenditure and hurdles, which has improved productivity in the industrial sector. Computer and TV games have become bigger and better. There was a time when "Super Mario" was considered the best game.[6] As in other sectors, in this sector too, AI has proved effective, cheap, and productive.

7.3.1 Jurisprudence Analysis of Artificial Intelligence

AI is a challenge before the law because of its implications and the wide spectrum of its application. The law only covers and protects the interests of humans. Actually, AI is not a person in the eye of law. Secondly, AI is the brainchild of human efforts that works with the help of some program and software, which can be managed and regulated through various machines and robots. All laws are made with the intent to punish a person if he or she has committed any wrong or breach of law. All argument of law are settled with human beings—rather around the "person." Legal reasoning involves a great deal of interpretation of legal terms and predicates, many of which have meanings or definitions that are inherently indeterminate.[7] The application of all these terms and definitions are around the person and his or her liability under the civil and criminal law. There is no such law that punishes the activities of machines or monitor the act of machines. Activities through computers, mobiles, smart devices, machines, etc., are not covered under the preview of the law. The person who handles it may be responsible for the same if found guilty or having *mens rea* at the time of doing the particular act.

Under the civil law, there are various concepts of liability where the person is liable for his or her wrongs. In the civilized society, the conduct and behavior of a person can be managed and monitored through rules or legal principles set by the society. All people obey the same in order to maintain peace and ensure development in the society. The concept of the liability or responsibility of wrong or offences starts with the bad and good actions of an individual. If the

---

[6]Rounak (Jan 3, 2017). What is the importance of artificial intelligence in everyday life? available at https://globussoft.com/importance-of-artificial-intelligence/.

[7]Rissland, E. L. (1988). Artificial intelligence and legal reasoning: a discussion of the field & Gardner's book. *AI Mag.*, **9**(3):45–55.

individual does any wrongful or unlawful activity in the society, he or she may be liable for the wrong or damages. The whole theory of liability is based on the principle of duty, which is fixed by law. Every person is bound by some duty given by law. The point here is that all these duties are for people only. It means that if a person breaches his or her duty, then he or she may be liable for punishment by the law. This means that to cast liability of any wrong, there must be a "person." As far as AI is concerned, on whom can we cast duty? AI is neither a legal person nor an artificial person who comes under the purview of a "person." For more clarity, let's discuss the legal meaning of a "person."

The term "person" is derived from the Latin term *persona*, which means those who are recognized by law as capable of having legal rights and being bound by legal duties. As far as jurisprudence is concerned, the term "person" refers to a community of humans or an organization or a natural person on which law can apply for regulating the conduct or affairs of that person. The legal person has some rights, duties, and liabilities in the eye of law. In other words, a "person" means anyone who regulates his or her conduct as per norms or laws. Additionally, a natural person means one who acquired legal personhood on birth in the society.

According to Salmond, "A person is any being whom the law regards as capable of rights and duties. Any being that is so capable is a person, whether a human being or not, and no being that is not so capable is a person even though he be a man."[8] An analysis of Salmond's definition shows that a person means anyone who is able to enjoy the rights and obey the duties fixed by the law. Additionally, it is not necessary for the person to be a human. Any artificial person may be within the ambit of being defined as a "person."

As per Savigny, "The person as the subject or bearer of right."[9] This means anyone who is capable of enjoying rights with legal limitation is a person. Furthermore, Gray said that "a person is an entity to which rights and duties may be attributed."[10] Gray advocates the rights and duties of individuals. Anyone who has

---

[8]Salmond, J. W., Fitzgerald, P. J. (1966). *Salmond on Jurisprudence*, 12th ed., Sweet & Maxwell.
[9]Mahajan, V. D. (2009). *Jurisprudence & Legal Theory*, 5th ed., EBC, Lucknow.
[10]Ibid.

rights and duties or who enjoys rights as well as duties is within the meaning of "person."

On the basis of the above definitions, the term "person" can be classified under two broad categories:

- Natural person (human being)
- Artificial person or legal person or juristic person

We must take into consideration the fact that AI is neither a natural person nor an artificial person having some legal entity. Actually, AI is a program that can work through devices that are not visible. In other words, it is the invention of a person not having any legal or artificial identity. This means that AI is not covered under the purview of a person and hence it has no liabilities and rights under the law. AI is a program that can be managed and operated through someone else. Now the question is, who is responsible for the act through AI? The answer to this question is but obvious: either the person who invented the program or the person who operated the same is liable in the eye of law. Now the theory of *mens rea* comes into picture. The principle of *mens rea* is based on the maxim *Actus reus non facit reum nisi mens sit rea.* his means unless there is a guilty mind, the person is not liable in the eye of law. AI has no separate brain or identity and for this reason has no *mens rea*; it only works as per the will and wish of its maker or inventor. To determine the guilty intention of the person, one must judge or determine the act. In addition, one must determine the manner, intention, modes of action, and gravity of injury to hold anyone responsible under the criminal or civil law.

7.3.2 AI Technologies and Liability

Under the broad ambit of AI, various technologies have also been developed over the years. Given below are a few definitions of the different technologies developed over the years and their current market share.[11]

---

[11] 2017, The year ahead: artificial intelligence; the rise of the machines; report by Merrill Lynch – Bank of America, dated 09 December 2016.

- **Machine learning**: ML includes computer processes based on scientific representations using possibility to make conventions and can make calculations about similar data sets. If the data are wrongly set by the inventor or programmer, then how can AI be responsible under the law for loss or injury caused to an individual. It is the responsibility of the inventor or programmer to load accurate data on the machine. For example, if data are inaccurately fixed by the inventor, giving wrong results, and because of this a person suffers a loss, the inventor will be responsible under the law.
- **Cognitive computing**: This involves using large data sets with the goal to simulate the human thought process and predictive decisions. Training of the systems tends to utilize human creation. If the data are wrongly inserted, the human prediction will come wrong.
- **Deep learning**: This involves neural nets to make predictive analysis. The use of neural nets is what differentiates deep learning from cognitive computing. Deep learning also helps improve image and speech recognition. The process of deep learning and speech recognition is based on the program fixed in the machine. If the inventor does a mistake, the machine will definitely misbehave and produce misleading results.
- **Predictive application programming interfaces**: It basically uses AI to provide a predictive output when you have data sets.
- **Natural language processing**: NLP programs computers to understand written and spoken language, just like humans, along with reasoning and context, and finally produce speech and writing. Many ML companies use NLP for training on unstructured data.
- **Image recognition**: This helps recognize pictures and objects as humans, as well as patterns, in visually represented data, which may not be otherwise apparent.
- **Speech recognition**: This converts spoken language to data sets that can be processed by NLP.[12]

---

[12] All points refer from cited in Nishith Desai Associates. The future is here: artificial intelligence and robotics, May 2018, available at www.nishithdesai.com.

7.4 Rights, Duties, and Liabilities of the AI Inventor

The question of rights and duties arises when violation of law takes place. As per law, if you have rights, you are also bound by some duties, fixed by the general principle of law. Rights and duties are supplementary and complementary to each other. As per law, every legal and artificial person has some rights and duties to perform. The question may arise in case to fixed the duty on AI inventor. As far as the rights of the inventor are concerned, he or she may have ample rights, from pecuniary interest to other benefits, including patents, but when the question arises about the liability, said person may be exempted sometime due to inadequate application of law. Rights and duties go hand in hand. There is a close nexus between rights and duties, without duties, rights have no meaning and vice versa. Wesley Newcomb Hohfeld[13] analyzes the relationship between rights, duties, and immunity, as shown in Fig. 7.1.

One of the greatest hindrances to a clear understanding, an incisive statement, and a true solution to a legal problem frequently arises from the express or tacit assumption that all legal relations may be reduced to "rights" and "duties" and that these latter categories are, therefore, adequate for the purpose of analyzing even the most complex legal interests, such as trusts, options, escrows, "future" interests, corporate interests, etc.[14]

Jural opposites comprise the following:

- Right/No right
- Privilege/Duty
- Power/Disability
- Immunity/Liability

---

[13] A professor at Stanford University and later Yale University who wrote only a few articles before his premature death in 1918. His most famous article "Some Fundamental Legal Conceptions as Applied in Judicial Reasoning" became a canonical landmark in American jurisprudence.

[14] Lazarev, N. (2005). *Hohfeld's Analysis of Rights: An Essential Approach to a Conceptual and Practical Understanding of the Nature of Rights*, http://www.austlii.edu.au/au/journals/MurUEJL/2005/9.html.

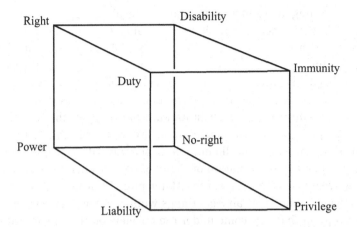

Figure 7.1 Showing the relationship between rights, duties, disability, privileges, liabilities, immunity, and power.

Jural correlatives are mainly made by:

- Right/Duty
- Privilege/No-Right
- Power/Liability
- Immunity/Disability

The analysis made by Hohfeld is quite complex and hard to understand. The reason behind this confusion is inaccuracy and ambiguity in the meaning of these terminologies. The terms "right" and "duty" have different meanings and may be used as per convenience of the parties. And, therefore, the interpretations and meanings of these two terminologies cause muddled analysis.[15]

We use the term "right" in different contexts; sometimes it may reflect power, immunity, and privilege and sometimes it simply refers to the rights guaranteed either by state or by law. The term "duty" has different meanings, too. It may be considered a moral duty, a legal duty, an ethical duty, etc. This means that rights and duties have different connotations, not allowing the law to cast and fix duty on the wrongdoer.

---

[15] Nyquist, C. (2002). Teaching Wesley Hohfeld's theory of legal relations. *J. Legal Educ.*, **52**:238–257.

According to Hohfeld, fundamental legal conceptions are *sui generis*, which means that all attempts aimed at creating a formal definition are not only dissatisfying but also useless.[16] It also means there is no logic between duties and rights because they are correlated. In simple words, if I enjoy some rights, this must be with some restrictions, and these restrictions are known as duties. The concept of duty means some limitation when enjoying the rights.

The principles of rights and duties are also applicable to the AI inventor. If the inventor did not follow his or her duty properly, this may amount to the violation of rights of others. Then the question may arise, what is the duty of the AI inventor? Basically, the inventor is legally bound to do all legal things while making any instrument or program that may come under the purview of AI. If the inventor did his or her job without any mala fide intention, then the program will be fruitful for others. This means it is the duty of the inventor to perform his or her task with utmost care and caution, which amounts to his or her duty. At the same time, the inventor has right to get monetary and other benefits generated through his or her invention.

A right means a legal claim of one person on another to act in a certain way. An AI inventor's right is to perform or make such programs or inventions that will work as per the intent and the same must be bona fide. An AI created for destruction or for the commission of crime is not within the meaning of the rights of a person. The AI inventor has a duty to perform his or her task as per the principle of law. It means the rights of an AI inventor lies in a duty to perform as per the law. In simple words, an AI inventor ought to behave in a certain way—in a legal way. The creation of AI is the right of the inventor; at the same time legal duty is imposed on the inventor. The sphere of my legal liberty is that sphere of activity within which the law is content to leave me alone. Privileges may be accompanied by rights that impose duties on other people not to interfere. However, privileges can sometimes exist without the existence of a right.[17]

---

[16]Chugh, D., Sharma, S. Hohfeld's analysis of legal rights, available at http://ijlljs.in/wp-content/uploads/2014/10/article.

[17]Hohfeld, W. (1913). Some fundamental legal conceptions as applied in judicial reasoning. *23 Yale Law J.*, **16**:28–59, http://www.law.harvard.edu/faculty/cdonahue/courses/prop/mat/Hohfeld.pdf.

It is the duty of the AI inventor to make an AI machine that is:

- Accurate
- Cost effective
- A long-lasting working module
- Updated and new
- Transparent in terms of operation
- Without any hidden setting
- Intelligent and smart
- Deep minded
- Secured (not hackable by others)

The AI inventor/maker has the following rights:

(1) To get royalty
(2) To get a patent
(3) To get a share in profit
(4) To not disclose the program
(5) To expand the business

As discussed above, every right and duty of the AI inventor correspond to each other—they are two sides of the same coin. The duty of the AI inventor is an integral feeling of obligation toward someone. If the AI inventor has the right to get a share in the profits, it is binding upon him or her to act as per the legal norms. This implies that if X, an inventor of AI, enjoys a right against Y, then Y is duty bound to respect this right. Rights, in the strict sense, can, therefore, be held to be benefits that are derived from duties imposed upon others.[18] If the inventor of AI has perfect rights, he or she has the perfect duties fixed by the law. These perfect rights and duties are not only recognized by state law but also come under the purview of strict legal action fixed by the law of the state. To enforce rights and duties, the state has the right to take strict and legal action, including force, if necessary, to implement these in a legal manner.

The second part of the above analysis is about liberty, which means the exercise of rights without the intervention of law.

---

[18]An essay on 'The Hohfeldian Analysis of Rights Philosophy Essay,' available at https://www.ukessays.com/essays/philosophy/the-hohfeldian-analysis-of-rights-philosophy-essay.php#ftn8.

Basically, the AI inventor has the liberty to act as per his own choice. He or she can do anything without the interference of law subject to the protection of interests of others. Liberty certainly does not mean that one can do anything that is against the law or morality. The existence or enjoyment of liberty is limited by the protection of interests of others. A classic example of perfect liberty is one where no one has any exclusive right to prevent the occurrence of a given act.

The term "liberty" has a wide meaning. For example, it does not mean the AI inventor can interfere with or violate the liberty or rights of others. In other words, the AI inventor cannot cause injury or harm through his or her invention to others. An AI inventor is at liberty to earn a living in his or her own way, provided the inventor does not violate some law prohibiting him or her from doing so and provided he or she does not infringe upon the rights of other people. This involves the liberty to deal with other persons who were willing to deal with him or her. This liberty is a right recognized by law; its correlative is the general duty of each person not to prevent the free exercise of this liberty or right, except so far as his or her own liberty of action may justify him or her in so doing. But a person's liberty or right to deal with others is nugatory unless they are at liberty to deal with him or her if they choose to do so. Any interference with their liberty to deal with said person affects him or her.[19] The term "liberty" has a wide meaning that cannot be enjoyed at the cost of others' freedom of rights. Liberty, therefore, is the existence of the "unconstraint" permitted under law.

7.4.1 Ethical Responsibilities

AI has a great impact on the day-to-day life of humans. And an increasing number of people are more and more dependent on AI inventions. With this have come new challenges not only before the inventor but also before the users. The right to privacy and other individual rights will be disturbed if nonethical practices are adopted by inventors. The use of machines and computer-related programs has led to a massive increase in the hostile approach

---

[19] Quinn vs. Leatham (1901) UKHL 2.

of some inventors. Actually, these machines and programs do not cause any direct or physical injury to individuals or users. But various rights have been violated through these inventions. The moral responsibility of this violation lies with their inventors. And so comes into question the moral status of inventors and their machines. Any inventor who makes as program with mala fide intention may cause harm to users.

AI differs from humans in many aspects. Humans have emotions, sentiments, and reactions, unlike AI. For this reason, it is hard to define and discuss the ethical responsibility of AI. AI is often more intelligent than humans, but it does not have emotions and the mind to discern the bad from the good. Due to this, the ethical or moral responsibility lies with the inventor of AI to manage and perform the task of invention while adhering to a high moral standard.

In near the future, questions may arise about the ethical behavior of machines or AI. These machines have no logic or practical experience of dealing with problems that arise on field. They work simply as per the programs installed on them by the inventors. Also, the maker of a program may not be aware of the practical difficulties of the person using the program. If the AI or software does not act as per the wish of the people, then it may cause ethical violation of principles of law.

Ethics or ethical practice is based on the accountability of persons. If the machines misbehave as per the program, the credibility of accountability may come into question. Practically speaking, the machines or AI is not accountable to anyone, simply because it is a machine. The fundamental principle of ethics is to maintain the honor and dignity of individuals for securing friendly relations between two. In other words, ethical practice indicates the establishment of brotherhood among people. And as machines do not have any emotions and sentiments, it is difficult to build friendship and brotherhood between machine and man. From the ethical point of view, it is the sole responsibility of the maker to create and manage ethical programs for machines. The responsibility of ethical practice lies with the inventors, not with the machines.

Establishing an ethical relationship between the machine or AI and the human being is the sole responsibility of the inventor. If his or her act is not bona fide or ethical, how can the AI be responsible for the misbehavior? The inventor is required to follow certain principles and notions of decorum when designing and monitoring machines or AI as an ethical responsibility. The breach of ethical duty by the inventor can lead to sanctions against him or her. Additionally, it may be considered professional misconduct, leading to punishment.

The ethical liability or responsibility is with the inventor or maker of the AI or program. This gives rise to the challenge of machine, its designer, and ethics. If we hold the maker or designer of the machine liable, it may cause injustice to him or her. The inventor or designer makes the AI and other software as per the need, demand, and predictable work of the person or the company. If the AI or program is been manipulated in some way by the inventor, then any related damage or harm is the sole responsibility of the inventor, not the user, of the machine.

Another important social criterion for dealing with organizations is being able to find the person responsible for getting something done. When an AI system fails at its assigned task, who takes the blame? The programmers? The end users? Modern bureaucrats often take refuge in established procedures that distribute responsibility so widely that no one person can be identified to blame for the catastrophes that result.[20] Ethically, it is hard to define what the ethical responsibility of an AI inventor or programmer is. There is no clear-cut definition of the ethical responsibility of the AI inventor and programmer, and there may be diverse opinions regarding this.

On the basis of various assumptions and possibilities, one can say that the inventor or programmer of AI is ethically bound to obey all ethical principles expected from professionals. The list is exhaustive, but some are as follows:

- The programmer or inventor of AI must know all the possibilities and the related disadvantages for the society and laws.
- They must be aware of the AI's negative and positive implications if they accept or adopt the same.

---

[20] Bostrom, N., Yudkowsky, E. (2011). The ethics of artificial intelligence. In *Cambridge Handbook of Artificial Intelligence*, Ramsey, W., Frankish, K., eds., Cambridge University Press.

- They must know the probable loss if the AI goes beyond specified limits.
- The constitutional and other rights of the people must be protected through the program.
- If the AI goes beyond human supervision, it should not be used.
- If the AI or program is ambiguous, it should not come into picture.
- Confidentiality is crucial during the designing of the AI or program The maker must take care of the confidentiality of the program. If the same is disclosed to common people or a hacker, it may be used adversely.
- The program—its use, process, and technical issues—must be disclosed to the employees or persons directly or indirectly using the same.

AI is electronic devices or programs that can be operated and managed by humans. These machines or robots have no morality or ethical values as they are an accumulation of some instruments that can be operated as per the choice of the person operating them. Therefore, you cannot expect morality from machines or robots. The root of morality and legal principles must be embedded in the mind and heart of the maker of this program. For example, if I feed biased data into the machines, the results may ultimately be biased because the machine is not in a position to determine the emotions, anger, rights, and involvement of persons and positive and negative impact on society and persons.

Machines are not humans and cannot give attention, kindness, love, and affection to humans or react against them. Machines cannot determine right and wrong.

Humans works on values set by the law or by society. Machines are not designed to learn human values. Recently, an attempt has been made to create such a machine, but results are yet to come. IBM researchers are collaborating with Massachusetts Institute of Technology to help AI systems understand human values, by converting these values into engineering terms. Stuart Russel pioneered a helpful idea known as the value alignment principle, which can help in this area. "Highly autonomous AI systems should be designed so that their goals and behaviours can be assured to align with human values throughout their operation."[21]

7.4.2 Criminal, Civil, and Constitutional Responsibility of the Inventor and AI

As discussed, the AI inventor is liable both ethically and legally. The inventor of the AI or program is held responsible for any loss or injury caused to any person due to the use of the AI or program. As per the constitutional law, every individual has fundamental rights, abridgement or violation of which leads to legal action in the court of law. These fundamental rights—rather human rights— are provided at the globe level, where every person has the right to enjoy the same without any kind of discrimination. The violation of these fundamental rights by the person or states leads to legal action, which may result in getting compensation from or giving punishment to the wrongdoer.

The constitution of any country refers to the fundamental law of the land, which provides overall protection to every citizen against any kind of discrimination and hate. The constitution provides a series of protection for the overall development of every individual, without any kind of discrimination. The structure of rights assured under the constitution does not discriminate on the grounds of sex, age, place of birth, color, etc. It means all people are equal in the eye of law. All humans have an equal status and their dignity must be protected.

The basic human rights include the principle of equality before the law and equal protection of law.[22] It means all the people are equal in the eye of law and the state shall not make any kind of discrimination merely on the grounds of sex, religion, place of birth, color, and caste. The state is bound to provide equal treatment to everyone. On the same line of arguments, if a wrong is committed

---

[21]Thurai, A. (Jan 15, 2019). It is our responsibility to make sure our ai is ethical and moral, available at https://www.aitrends.com/ethics-and-social-issues/it-is-our-responsibility-to-make-sure-our-ai-is-ethical-and-moral/.

[22]See Article 2 of the Universal Declaration of Human Rights and Article 14 of the Constitution of India, 1950.

by an AI or through its programmer, the person shall be responsible and liable for compensation.

The number of crimes through the use of AI, the Internet, and various smart devices is increasing on a daily basis across the world. The very right to privacy of individuals is under threat because of AI and the Internet. All the private information of the person can be monitored by someone not even remotely concerned with that person. Through Internet access, a person can stalk every activity of the intended victim, threatening the expression, privacy, and safety of the person. The digital evolution has made the Internet easy to access, particularly for children and teenagers, who are not aware of the risk and challenges.

The right to privacy and the right to life are basic human rights that should not be curtailed on the basis of nationality, place of residence, sex, nationality, or ethnic origin, color, religion, language, or any other status, because these rights are inherent in nature, granted by nature to every individual. The purpose behind human rights is to recognize individual status for regulating each individual's existence in society. These are inherent rights of individuals, and no human being should be deprived of these rights by acts of any state. The purpose behind enacting various laws and treaties is to protect human rights throughout the globe, but the ground reality tells a different story: the rights are not the same for all people and societies. They differ from nation to nation and society to society.

India is a member of the International Covenant on Civil and Political Rights and the International Covenant on Economic, Social, and Cultural Rights adopted by the General Assembly of the United Nations on December 16, 1966. The right to privacy and the right to life of every individual is paramount for the state. That is why Article 21 of our Constitution protects the life of every individual. Our Constitution also adheres to Article 25 of the Declaration of Human Rights, recognized as the right to health for individual and family.

The Constitution of India guarantees the right to life, the right to privacy, and the right to personal liberty for every individual under Article 21. The Supreme Court has held that the right to live with human dignity, enshrined in Article 21, derives from the Directive Principles of State Policy.

7.5 Civil Remedies under the Law of Torts

The principle laid down in the law of torts relating to damage or injury caused through AI and other smart instruments has various nuances. The damage caused by AI must be tested on the various principles under the law of torts. The principle of law of torts is based on some postulations as well as some responsibilities of the wrongdoer. The remedy of damages or injury is also allocated according to its gravity and loss to sufferer. Fundamentally, the loss or injury confers liability under civil and criminal negligence or intentional torts.

Although the tort of negligence or intentional harm is of general application, it may occur in a number of cases, such as misuse of power, wrong programming, wrong data, and fault during the manufacture of instruments. Civil wrong includes failure on the part of a programmer or inventor of AI to discharge obligations that arise out of the contractual nature of the contractual relationship. It also includes failure of some duties toward society.

The fact is that tort law is a mixture of the principles of corrective justice and distributive justice. And in a situation of uncertainty and difficulty, a choice sometimes has to be made between the two approaches.[23] This means that every tort does not amount to a crime under the law. A failure on the part of the inventor or programmer has certain essential components that must be seen by the court at the time of deciding the case. These are as shown in Fig. 7.2.

When awarding damages to the suffer, the court must satisfy the ingredients mentioned in Fig. 7.2 and then act accordingly. If the plaintiff fails to prove these elements, then the court should not pass any order relating to the award of compensation. This principle is not based on the principle of contract between the inventor and the service; its existence is independent of the violation of the rights of the person who suffered the loss or injury.

---

[23] McFarlane vs. Tayside Health Board (2000) 2 AC 59.

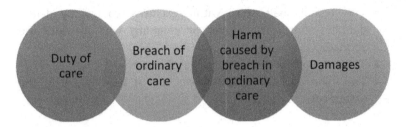

Figure 7.2 Showing the correlation between duty, breach of duty, harm, and damages to the person.

7.5.1 Principle of *Res Ipsa Loquitur*[24]

Generally, the burden of proof lies with the plaintiff. This rule has exception in case of failure of responsibility or wrongdoer case. If the negligent act itself is very obvious, the maxim "the thing speaks for itself" will apply and the inventor, the programmer, or the company will have to prove that the loss was not caused due to failure on its part.

Res ipsa loquitur is a good weapon in the hands of the opposite party. The person who suffers loss or injury may use it when there is insufficient evidence on record and he or she is confused about what caused the wrongful act but the circumstances clearly indicate that the inventor of the AI or the programmer is responsible for such failure.

It is hard to apply and establish even in civil jurisdiction the rule of *res ipsa loquitur*, which is not of universal application and has to be applied with extreme care and caution to the cases of professional mistake and wrong in particular fields. For attracting this maxim, the plaintiff must prove the following:

- The plaintiff did not contribute to the failure of any program or machine.

---

[24]A Latin word meaning "the thing speaks for itself," a doctrine of law that one is presumed to be negligent if he, she, or it had exclusive control of whatever caused the injury even though there is no specific evidence of an act of negligence and without negligence the accident would not have happened. Adapted from http://dictionary.law.com/Default.aspx?selected=1823.

- The default shall cause within exclusive control by defendant/inventors/programmer or company who provides the service.
- There is a prima facie case against the inventor.
- The injury is such that it shows that it was caused by the defendant.

7.5.2 Compensation under the Law of Tort

The word "compensation" has been derived from the Latin word *compensare*, which means "to compensate." The compensation usually involves repayment or indemnity (typically in the form of money) to a person who suffers loss, injury, or deficiency in service due to the act of the respondent.

So far as loss caused due to AI is concerned, there is no exact technique to fix tortuous liability except to provide compensation to the other party. This is the only method to give relief to a plaintiff in civil cases. The compensation can be awarded by civil as well as consumer courts. Compensation is a method to impose monetary prevention on the inventor or service provider company. Essentially, the compensation is awarded to maintain the status quo of the affected party or plaintiff.

Damages or compensation awarded under the law of tort for loss due to AI is not punishment to inventor or programmer or company. It is just an award for injury sustained and suffering by the act of another. The basic principle of right to life, which is guaranteed by the Constitution of India under Article 21, shall not be defeated by the other party. For saving as well as suffering of loss, the compensation shall be granted by the court.

The compensation under the law of tort is called "damages." These include various kinds, like nominal damages, substantial damages, contemptuous damages, exemplary damages, and aggravated damages. The cases of loss by AI and other programs and the related forms of compensation can be classified as damages and divided into two types:

- General damages
- Special damages

7.5.2.1 General damages

These include nonpecuniary loss (like the pain, injury, or sufferings of the plaintiff). They can be calculated as per the general term of human prudence, which is based on:

- The pain and suffering
- Loss of amenities
- Loss of anticipation of life

7.5.2.2 Special damages

These can be calculated on the basis of the actual monetary loss to the plaintiff. This is based on the assumption of:

- Loss of earning
- Cost of instruments
- Loss of future earning

7.5.2.3 Common principles for the contemplation of damages

These principles take into account:

- Loss of earning from the date of injury till the disposal of trial
- Other material loss, including nonfinancial loss
- Physical, mental, or physiological damage, injury, pain, or suffering
- Inconvenience, hardship, discomfort, disappointment, frustration, or mental stress
- Damage or loss of expectation of life
- Nature of damage or loss caused to the plaintiff due to the act of the defendant
- Cost of litigation in the court, including court fees, lawyer fees, and other expenses incurred due to such litigation
- Cost of mental suffering, which can be calculated on the basis of suffering to the plaintiff
- Loss incurred due to physical or mental disability
- Loss or injury or suffering caused to any dependent on the plaintiff

7.5.2.4 Standard principles for granting any damages

These principles take into account the following points:

- The loss claimed by the plaintiff must be reasonable and according to provision of law.
- The loss of suffering must have been caused by the defendant's act. The plaintiff must not have contributed in the act or suffering.
- The personal injury must have been caused due to the defendant's breach of duty.
- The amount claimed must be reasonable.
- The plaintiff should have actually suffered the loss or damage.
- The amount must be reduced if there is contributory negligence on the part of the plaintiff.
- The amount or claim should be rejected if the plaintiff files any false or fictitious lawsuit against doctor.
- The quantum of amount must be calculated if death has occurred due to the misuse of AI or any program.
- The calculation must be based on the facts and circumstances of each case.

The method of calculation of the compensation amount for loss caused due to the misuse of AI and other smart devices is based on various associated things. The compensation to parties is not measured by a golden scale because the loss or suffering, including injury, sustained to the mind, reputation, and person cannot really be reimbursed in terms of money. Money is a remedy to compensate monetary loss to the person. Suffering and sorrow of many kinds cannot be compensated in terms of money.

It is a settled principle of law that if a person acts negligently when performing his or her duty and causes any injury or loss to other person, he or she must be liable to pay compensation, which is calculated after taking all the circumstances into account. The basic aim of providing compensation is for the plaintiff to recover from the financial loss. Though the way of calculating damages is imperfect to a certain extent, there is no other way to compensate the person for his or her loss and suffering.

7.6 Liability under Criminal Law

Criminal law is only applicable to the person or individual who has directly or indirectly caused harm or injury to another. As far as AI and other instruments are concerned, there is no direct penal provision that punishes machines or robots. But if injury is caused by using any instrument, device, or weapon, then the person who used these may be responsible for the same under the law of crime. Actually, the wrongdoer under criminal law is the person who makes a device or program with mala fide intention—an aim to cause harm or loss to others.

7.7 Challenges Ahead

The Constitution of India and various civil and criminal laws provide overall protection to the individual, without any kind of discrimination. But merely providing protection is not enough to protect the rights of individuals. With further development of science and technology, the use of AI will increase in the near future, and it is necessary to provide better protection to individuals against harm from the misuse of AI and other software.

It will be remiss not to mention here that the courts are still in a dilemma where the legality and status of AI and other related instruments is concerned, largely because these are operated and handled by humans. The courts and legislature are still in the process of discussing which laws should be applicable for acts caused though AI and other smart devices. The Ministry of Industry and Commerce in India has accepted the challenges and benefits of AI in the industrial sector. But the other sectors are not yet ready to accept and monitor AI.

Recently, the importance of AI was accepted by the Election Commission, the Planning Commission, the Department of Income Tax, and the Unique Identification Authority of India, but they are still struggling with the challenges and shortcomings of handling and tackling the problems arising from the misuse of AI and other smart devises. AI is part of fields like agriculture, health, banking, national security, retail, and customer relationship. Errors often occur while handling AI. Now the time has come to think with an open mind about the working and practical implications of AI. Apart from the above, some of the important challenges before AI are as follows:

- There are no specific laws or rules that control the activities of AI and other devices. The existing laws are not sufficient for the same.
- The responsibility of the inventor or the program is still not fixed as per the existing provisions of law. If any person is found guilty of any activity, then who will be punished is a question with no answer as yet.
- Most people are not techno-friendly and not aware of the impact of AI.
- The public at large is still not open to accepting the existence of AI as it cannot be seen, touched, or perceived.

7.8 Conclusion

We are living in a civilized and scientifically well-developed world, with many facilities and amenities available at the click of a button. Human efforts have reduced due to the use of science and technology. People enjoy a hassle-free life while deriving maximum satisfaction possible. Various smart devices and instruments work tirelessly for human pleasure and luxury. In almost all sectors, robots and smart devices have taken the place of humans. Due to the use of technical gadgets, our economy is developing and we save money, but at the cost of human labor.

Gradually, people are coming to depend more and more on machines and smart devices, which has led to instances of violation of their rights and unnecessary interference in their personal lives. Many problems have arisen due to the misuse of these devices and AI. To top it, there is no appropriate legal jurisprudence on the subject that can measure the misuse or neglect and provide solutions. The existing laws and policies of the government are not adequate enough to provide complete protection against criminal acts or acts of negligence committed through AI.

References

1. Singh, A. (2017). *Contract and Specific Relief*, EBC publication.
2. Courville, A., Goodfellow, I., Bengio, Y. (2016). *Deep Learning*, MIT Press.
3. Reese, B. (2020). *The Fourth Age: Smart Robots, Conscious Computers, and the Future of Humanity*, Simon and Schuster.
4. MacMahon, B., Binchy, W. (2007). *Law of Torts*, 3rd ed., Tottel Publication.
5. Charniak, (2002). *Introduction to Artificial Intelligence*, 1st ed., Pearson Publication.
6. Chollet, F., Allaire, J. J. (2018). *Deep Learning with R*, Manning Publications.
7. Postema, G. (2001). *Philosophy and the Law of Torts*, Cambridge University Press.
8. Hallevy, G. (2014). *Liability for Crimes Involving Artificial Intelligence Systems*, Springer.
9. Ghormade, V. (2014). *Lecture on Jurisprudence & Legal Theory*, 2nd ed., Hind Law House.
10. Gaur, K. D. (2009). *Text Book on The Indian Penal Code*, 4th ed., Universal Law Publication.
11. Ashley, K. D. (2017). *Artificial Intelligence and Legal Analytics: New Tools for Law Practice in the Digital Age*, Cambridge University Press.
12. Tondon, M. P., Anand, V. K. (2017). *International Law and Human Rights*, 18th ed., Allahabad Law Agency.
13. Corrales, M., Fenwick, M., Forgó, N. (2018). *Robotics, AI and the Future of Law*, Springer.
14. Shapo, M. S. (2003). *Principles of Tort Law*, 2nd revised ed., Thomson/West Publication.
15. Nilsson, N. J. (2009). *The Quest for Artificial Intelligence: A History of Ideas and Achievements*, Cambridge University Press.
16. Bodenheimer, E. (2001). *Jurisprudence the Philosophy and Method of the Law*, 2nd Indian reprint, Universal Law Publication.
17. Jain, M. P. (2018). *Indian Constitutional Law*, 8th ed., LexisNexis.

18. Norvig, P., Russell, S. J. (2009). *Artificial Intelligence: A Modern Approach*, 3rd ed., Prentice Hall.

19. Joshi, P. (2017). *Artificial Intelligence with Python*, 1st ed., Packt Publishing.

20. Heuston, R. F. V. (1996). *Salmond on the Law of Torts*, 21st illustrated ed., Sweet & Maxwell.

21. Draft Committee (1948). Universal Declaration of Human Rights.

Chapter 8

Jurisprudential Approach to Artificial Intelligence and Legal Reasoning

Arup Poddar

School of Technology Law and Development, WB National University of Juridical Sciences, Kolkata, India
arup.poddar@nujs.edu

Artificial intelligence (AI) is playing an ever-increasing role in everyday human life. In artificial intelligence, there is an absence of the application of human mind in devising intelligence. For example, addition or multiplication, which can be done by human beings, can also be done by a small calculator. After a human being has fed data into the calculator for addition or multiplication, it is the calculator that by its own intelligence reaches a scientifically appropriate result, which is nothing but an example of artificial intelligence. Though the calculator is an example of low-level artificial intelligence, in the present society, one can find a number of areas where high-level, multidimensional, multifaceted, and competent artificial intelligence is being used, for example, in robotic surgery, in supercomputers for space programs, in missile technology software, in automobile software, in flight software, and in Google Maps for reaching a destination or tracking a person's location. Another example of high-level artificial intelligence can be witnessed in IBM's

Artificial Intelligence and the Fourth Industrial Revolution
Edited by Utpal Chakraborty, Amit Banerjee, Jayanta Kumar Saha, Niloy Sarkar, and Chinmay Chakraborty
Copyright © 2022 Jenny Stanford Publishing Pte. Ltd.
ISBN 978-981-4800-79-2 (Paperback), 978-1-003-15974-2 (eBook)
www.jennystanford.com

Watson software program, which for the first time has stepped into the field of law for providing legal reasoning, with almost accurate answers to legal questions within a few seconds while the most intelligent person would take hours to provide almost-accurate answers. Artificial intelligence in the field of legal reasoning can be applied in multidimensional ways, for example, reasoning with rules, reasoning with cases, reasoning with precedence, and reasoning with the opponent's argument/counterargument. In this regard, the jurisprudence of artificial intelligence and legal reasoning would be a state of affairs where artificial intelligence will take the lead and play a competent roles as a judge, an advocates, a teacher, a law officer, a legal consultant, etc. This approach will definitely lead to unemployment issues in many sectors of law. For example, artificial intelligence software will give training to law school students on how to operate the software for reaching conclusions with the help of logic and ethics [1]. So people will no longer be required as teachers. There will be the possibility of a client contacting the artificial intelligence software company to hire an artificial advocate to take the case before the court of law and the judges will be relying on the argument developed by the artificial intelligence software and will decide the case accordingly; so people will no longer be required as advocates. A similar thing has happened where the traditional cameras and films are concerned, where they have almost disappeared because of the explosion of smartphones and digital cameras. The present chapter will analyze the scope of artificial intelligence in legal reasoning, discussing the merits and demerits of using artificial intelligence software.

8.1 Introduction

Serious consideration is being given to the adoption of AI in industries and for professional work, keeping in mind the potency of AI, which can work better than people in a lot of areas and can replace a substantial population of the workforce with the ability to work harder, longer, and more accurately. Joblessness would definitely be evidenced from the promotion and adoption of AI in various spheres of life and may be a risk to the very existence of human society [2]. It is believed that attorneys trained by law schools to act as superefficient lawyers to contest cases for their clients can very well be replaced by AI, which is nothing but an advanced technology with the ability to compute data helpful in generating a comprehensive argument that will not only be convincing but also be accepted by the forum before which the AI will advance its arguments. Therefore, with the help of an increasing database of national and international legislations, including the decided cases available within the software database of the AI, we will show that disruptive innovation has taken over specific statutory provisions with the help of new technologies.

It is also true that in legal academia, there is this discussion on how AI can take over the legal profession, making attorneys jobless. In fact, little is known to lawyers and students of law schools about the innovativeness and distinctiveness of AI software that will have the potential to generate arguments with the help of sufficient legal materials within a few minutes or even seconds, in comparison with a human lawyer, who may take days or even months depending on the quantum of cases filed on his or her behalf. It remains a matter of debate how efficiently AI can resolve legal disputes with the help of arguments generated by a database on the legal materials already available within its software along with all the recent updates. Because AI will compute the legal materials already available within itself on the basis of the question raised before it, the application of software to the legal materials will reflect the argument. Only the future can tell us whether the society is ready to adopt AI in legal reasoning, replacing the human workforce.

An internal combination of various subjects, such as biology, neuroscience, linguistics, economics, mathematics, computer science, philosophy, law, and psychology, is integral to any AI software that can generate legal argument with the help of legal reasoning. The view of John McCarthy, the person who for the first time brought up the idea of AI with the help of Michael Mills, is worth mentioning, and he stated that AI is a conglomeration of academia involving commercial work that involves engineering and science for bringing out an intelligent machine.

As per this view, it is also true that in practice of law, there is no implementation of image recognition or robotics but only legal reasoning developed from legal materials already fed into the intelligent machine and available in the form of a database. It is interesting to note here that AI experts, such as Russell and Norvig, provide different views in contrast to the views already explained above. According to them, AI systems are:

- Systems that act like humans
- Systems that act rationally
- Systems that think rationally
- Systems that think like humans

Is there any possibility that AI can solve legal problems? Can a computer application provide appropriate solution to any legal problem? Is it possible to enhance the software program of the computer so that legal reasoning can be developed in real time? How can AI help a real lawyer with a real solution? It is true that an AI computer application is not limited to typing in the text and editing the same. Nowadays, computer applications can automatically search the case law and legislation and generate specific answers. Therefore, the computer can provide a solution to any legal problem.

A computer application works on the data already fed into the system. Therefore, there is a fair possibility that a computer with higher specification can solve a legal problem. An AI system demands automated functioning based on the queries raised before it (see Ref. [3]). The specifically designed computer program can provide sufficient information to a lawyer, as well as all relevant case laws, articles, judgments, and statutory provisions, as a part of the solution to the given legal problem. A computer system with higher specifications can provide such a decision, along with the argument and the relevant legal materials online and offline, provided the data are already available in its super hard disk or in the cloud computing system [4]. At the same time, the computer can give a lecture on a specific topic with all the relevant legal materials within a specific timeframe as already requested of such program software. These are all good example of the application of AI in the legal field.

Any legal question provided to a computer program will definitely inform the search system programmed within the computer application and will try to provide the specific answers automatically while it surfs all the online and offline database available to its own cloud computing system. It is not possible for a person to remember every relevant legal point at the time of presenting a case before the court, whereas an AI computer application can provide all the relevant legal materials and present these before the court without missing any important points, an appropriate use of AI. It is not proposed that the computer application act as a judge to a dispute, but it can definitely provide relevant legal materials along with a checklist that a lawyer generally takes into consideration prior to presenting the case before the judge. This is an example of computer programming based on a computational model.

It is an accepted fact that a large mechanical part of the legal field is to do with the application and validity of procedural rules and it would be wise for such an application to be included within the computer application to ensure no procedural mistakes are committed. Subjects like international law, civil procedure, and conflict of laws/private international law are peppered with complex mechanical rules and structures. Accordingly, it will always be wise that a computer application with its own AI be involved in selecting relevant procedure and provisions thereof, which will be without any mistake and will be convincing in nature for a debate or for legal research. The seriousness of AI can be understood from the fact that it cannot overlook any relevant provisions, including exceptions or deceitful exception, which human beings can definitely overlook. Overlooking a provision might result in a negative impact on the case, such as application or nonapplication of domestic or substantive procedural law or at the international level foreign substantive or procedural laws. Computers are created by the human brain. Therefore, it cannot be considered that computers are more intelligent than human beings. Accordingly, it is possible that a human being may be physically slower than the AI but with the inbuilt creativity of the human brain, human creativity will be much better than that of the AI. But it is also true that even if the comment (answer/argument) developed by a supercomputer can't be at par with that developed by a human brain, with its creativity, the kind of human errors that a person will commit while doing a repetitive work will never be committed by a computer. Therefore, as an author, I feel that computer application as a part of AI can play a vital role in investigating and diagnosing the real legal problem and providing a solution thereof with the help of a checklist. It is also possible that the legal reasoning developed by the AI will be far better than human reasoning because there will be no mistakes in choosing relevant provisions and there will be no oversight or omission of important provisions, which may be the turning point of the case, leading to a win.

8.2 History and Origin of Artificial Intelligence and Law

AI is not of recent origin. It has been a part of society for many decades. One can find traces of AI in ancient Greek culture. In 1950, Alan Turing turned AI from his fiction work into a reality and stated "the idea of machines that think." It is also true that the idea of AI did not take off as expected during the 1950s. McCarthy, a computer scientist, in the year 1956 for the first time made a significant reference to AI. Thereafter, an AI laboratory was successfully established in the year 1959 at the Massachusetts Institute of Technology (MIT). The concept of AI attracted a lot of criticism, particularly during the 1960s and the 1970s, in many areas, such as AI progression, common sense reasoning, and logical program. However, all these criticisms came to an end when humans were introduced to the personal computer, which started solving many problems dealt with earlier by human beings [5].

Intelligent behavior is something that depends on the quantum of knowledge, and such behavior will be meticulous when there is a database of detailed knowledge. Researchers found this to be true in the 1980s, during the developmental phase of computer programming. In legal reasoning, computer programming must be based on detailed knowledge of law, cases, and international instruments already provided to the computer program, on the basis of which, the computer can provide comprehensive reasoning on law. Intelligent behavior has been adopted by most of the corporate houses and companies, not only to navigate behavior of human beings but also to provide a checklist of the daily business activities. To build comprehensive, articulate, and meticulous computer programming, which could learn, translate, and interpret the provisions, articles, cases, and general statements, more funding is required to be invested in the computer projects around the world.

We all know the story of Deep Blue, a computer program of IBM that became the first one to defeat the world champion Gary Kasparov in a game of chess. This was the greatest achievement at the time because the computer was able to compute 200 million moves per second. This was nothing but the initial stage of AI as it was not possible for any human being to think of so many moves per second. Since the project Deep Blue was a grand success for IBM, it ventured into higher challenges and introduced another level of AI with Watson, a computer program that can understand the natural language provided to it with a high-level combination of sophisticated software and hardware that not only delivers answers with precision but also provides appropriate justification with supporting evidence. Accordingly, the creation of Watson itself became a risk to the former champion Deep Blue.

A legal research tool has been developed by Watson, named ROSS, which can improve the performance of law firms beyond human imagination. When ROSS is asked a question, it goes through a billion text documents per second and then produces its own research statement within a few minutes. A very important aspect of ROSS is that it learns from feedback, just like the human brain. It is interesting to note here that Watson and ROSS learn laws, understanding the data principles and judgment of cases. Accordingly, they do not generate a statement merely out of keywords but rather perform deep research of the documents available in the database. Compared to Watson, ROSS is a more technology-oriented, sophisticated computer software program. ROSS was so valuable that in the year 2016 BakerHostetler, one of the biggest law firms in the United States, hired ROSS for performing meticulous research on bankruptcy law practice in New York. From the year 2016, ROSS became the first robot lawyer in the United States. After the tremendous success of BakerHostetler, other big law firms, such as Latham & Watkins and Simpson Thacher & Bartlett, also hired ROSS for facilitating their legal research. These are examples of AI in developing legal reasoning, but the amazing thing is that these machines learn by themselves and adopt more autonomy in devising answers to any queries through algorithm programs that they already possess, widening the scope of these algorithm programs to answer any given questions accurately and convincingly.

The publication entitled "Computing Machinery and Intelligence," which first came out in 1950 and was written by Turing, was the first step in realizing the importance of AI. The experiment mentioned in this publication subsequently became famous as the "Turing test" and came to be known for exposing the fact that a computer has a satisfactory level of intelligence. The level of intelligence can be considered sufficient when the observer finds that there is no difference between the answers given by computer intelligence and those by human intelligence. For example, if when a judge asks a computer and a human lawyer the same question and the answer given by the computer is not less than the answer given by the human lawyer, then this test is successful.

Surprisingly, the Turing test is not only highly appreciated but also severely criticized. It is a well-known fact that just like IBM's Deep Blue won the chess game in the 1996 against the best chess player of the time, a natural human being, in the year 2011, in the US trivia gameshow Jeopardy, it was IBM's Watson that won against two former winners, who were again human beings.

Initially, a low range of AI was introduced by the Turing test, which could include natural language processing, thinking, and learning.

The criticism of the Turing test is twofold: (i) that the comparison includes nonintelligent human behavior and (ii) that it does not include nonintelligent human behavior.

8.3 Types of Nonhuman Computational Capabilities

Many automated tests have been designed to understand and determine superintelligent nonhuman computational abilities:

- C-tests, or comprehension tests: These examine the comprehension capabilities, which are considered to be a main component of intelligence and required to be utilized to generate information when new data are fed into the system.
- Universal anytime intelligence tests: These assess the intelligence of any present or future artificial or biological system.
- The Winograd schema challenge: Levesque Hector, a professor of computer science at the University of Toronto, developed this test based on linguistic antecedents or multiple-choice questions prepared to gain information with the help of common sense, preliminary knowledge, and spiral and interpersonal skills.
- The logic theorist system: This system can impersonate the problem-solving skills of human beings and ensure high-order intellectual processes. This was an experiment by Alain Newell and Herb Simon.
- The Lovelace 2.0 test: This test was developed by Selmer Bringsjord in the year 2001 and perfected in 2014 by a Georgia tech professor Mark Riedl, who observed that there are works of art that require sufficient intelligence to carry out such art work.

There was another criticism of the Turing test, that of human misidentification—that it is not unusual for a human being to be misidentified as a machine. For example, an answer a judge wishes to have may or may not be generated by AI. AI as a discipline was for the first time developed in 1955 by the Dartmouth summer research project. McCarthy and Marvin Minsky were the participants in this project.

As per records, it can be clearly established that the first person to discover the term "artificial intelligence" is none other than McCarthy, an American computer scientist and cognitive scientist as well and one of the founders of the discipline of AI. The paper "Ascribing Mental Qualities to Machines" was published by McCarthy in 1979, in which he asserted that machines will have beliefs as they are samples of thermostats and because they will have beliefs, most machines will show the capability of solving problems.

Another American cognitive scientist in the field of AI, Minsky also developed the constructive AI doctrine. It was Minsky who stated that there is no fundamental difference between the human mind and computers. Minsky developed robotic arms and grippers, the first electronic learning and computer vision system. In *Perceptrons*, a book published in 1969 by Minsky and Seymour Papet, Minsky asserted that he felt he was prevented from developmental research on AI networks. Other than this, Minsky was an active researcher on the symbolic approach and the intelligence of human beings compared to that of AI. The positive thinking that machines will have humanlike intelligence capabilities was asserted by Minsky [6].

One of the earlier discussions on AI relating to law was presented in an article in the year 1970 by Bruce G. Buchanan and Thomas E. Headrick. The authors emphasized the fact that if AI were to develop legal reasoning at par with human reasoning, then AI will have to mirror or copy a person's reasoning process, which means that AI will have to mimic the analytical reasoning of lawyers to reach a legal solution. The same idea is inculcated in law students with regard to legal research and writing skills, which are taught to law school students throughout the world.

"The Study of Cognitive Processes Using the Conceptual Frameworks and Tools of Computer Science" is another seminal article that was published in 1990 on the same topic. As per this article, AI will have to develop its own understanding and synchronization of legal reasoning. The article reviewed the process of AI and how this intelligence can be used to practice law. The article says that if the system understands the model legal argument and develops its own argument to a given legal question, then AI can play a vital role in the practice of law [7]. This system is somewhat complicated, for example, in the first phase, AI will have to understand several types of knowledge related to law and arguments. In the second phase, it will have to develop reasoning with the help of material on the cases decided earlier, apply the relevant rules and regulation, and develop the argument of facts and law as per the need of the client. In the third phase, a computer program will have to be developed where AI can send all these data to ensure they fulfill all the parameters mentioned above. In the end, AI's goals will be to compile the available materials, for example, mixing various modes of legal reasoning; formulate explanations and arguments; and make changes in the explanation and argument while accommodating changes in the existing laws and newly decided cases [8].

For AI to impersonate the reasoning practiced by a human lawyer, it must be able to argue, research, think to determine relevant laws and facts, and apply previously decided cases, the way a human lawyer does. AI will have to be as adept as the human lawyer to change its course of argument depending on how the argument is panning out. AI has one benefit: since it makes billions of moves based on the texts available within its system, it is inevitable that AI will have to come with specific answers that a human lawyer may take a lot of time to come up with.

Once AI has understood using its own intelligence how a human lawyer uses his or her analytical ability on the basis of the given material, it is not a mandate for the AI to follow all those techniques that a human lawyer does. Once the AI has understanding like that of a natural lawyer, the concerned AI will not have to think like a human, act like a human, or even reason like a human with a human perception since the aim of the AI is to reach conclusive reasoning similar to that of the human lawyer or better than that.

There is another term available related to AI, "soft AI," which can attempt to impersonate human intelligence by devising the final result, but for that it will not adopt the human process. Soft AI also takes into account the fact that we have a limited understanding of the abilities associated with the human brain. Accordingly, without an understanding of the forces available in the human brain, no direct AI can work using the process as adopted by the human brain. It is also a well-established fact that a law firm will not hire AI for law practice where the software will be familiar with the issue, rule, application, and conclusion method or understand the reasoning in criminal law, or be able to pass any competitive examination. Instead, the law firm will utilize the AI system for arriving at the comprehensive conclusion/result and not for the knowledge already available within the system.

It is important to mention here that the computer scientist will develop the AI with a mandate that the AI must understand how to develop legal reasoning through a legal analysis of the existing rules and regulation and case laws or can reach a conclusion in collaboration with all the items [9]. However, developing legal reasoning at par with the skill of the human lawyer is always considered to be the higher end. If that is not the case, then there will be scope for recurring errors, which will damage the case. Those who are good as technical lawyers can develop the AI system, because developing the AI for legal reasoning requires multidisciplinary activity, which will involve legal data experts, lawyers, engineers, and users.

8.4 Introduction of Artificial Intelligence to Law

It is clear from the above discussion that AI is a branch of computer science research based on computational modeling behavior (the simulation of behavior/character by a computer program is called computational modeling behavior). If this behavior were the same as that of a human being, it will be called intelligence [10].

The aim is to develop an AI computer program that performs or simulates legal reasoning as a part of legal research as displayed by a human lawyer. For example, if a description of a legal problem is given as an input to the AI, then the machine will provide output by way of a solution to such a problem with meticulous explanation. A legal problem is always solved on the basis of the justification developed by legal reasoning not only for the present problem but also for futuristic problems yet to be faced. It is a simple rule that a decision without arguments is less important than a decision supported by arguments and justification. Accordingly, a computer program that successfully collaborates between argument and explanation is the key area of AI research for developing legal reasoning.

Under the computer program, generally, a computational model is described as one program that will be considered as input and another program that will be considered as output. There will be intervening steps—an algorithm—that will determine whether the output matches the input program or not. There can be three types of questions that can be put up for understanding the input/output behavior of AI:

- Can the problem be solved by understanding the input and the information used for analyzing the problem systematically? This will be answered with the help of knowledge representation and search.
- How can the search for a solution be regulated in terms of relevance and efficiency? This is related to the inference control mechanism.
- How can the program learn by its own errors and successes in order to improve its performance? This is related to further learning by AI.

A few materials, such as natural language texts, relevant cases, and legal rules, can be given as input to the AI and law program. It is important to note here that the problems are demonstrated in the AI as a specialized representation, for example, answer to a specific question generated by the expert system on generation of various dimensions as available on a given problem, and those directions may help to develop convincing legal reasoning. This will help to strengthen one's own views and weaken the opposite side's claim. In developing legal reasoning, AI can take the help of cases that have been decided earlier, along with the relevant rules and regulation, including all other points that the opponent has not mentioned in its plea. The combination of the law program and AI will have to focus on systematic research for a solution. For example, when searching the text for crime, the program must invariably search for relevant provisions of the criminal law.

It is true that in theory and practice the problem-solving behavior of AI and law programs can be based on performance and relevance factors, which is quite similar to a human lawyer solving a legal problem, for example, considering the coverage of relevant provisions, accuracy, precedence, and explanation. The AI and law program can also take the Turing test, for example, a human judge will be entitled to interact with the AI and law program and will ask different questions and will receive specific answers with explanations and reasoning. It should be the duty of the judge to accept the best possible legal reasoning with convincing justification to decide the case, whether such reasoning is coming from a human or AI.

Now, coming to the algorithm of an AI and law program, the intervening set of steps that ultimately help to convert the input material into different natural language texts and fix the connection with the relevant provisions of statutory law along with finding a database on similarly decided cases will give a viewpoint of how to generate legal reasoning. Therefore, the intervening set of steps (after material has been given as input) will have to go with the relevant legal materials that will help in explaining the output with reasoning and justification. Accordingly, the input/output behavior examines the process of the reasoning model in operation. The AI system will be considered successful when the legal reasoning output becomes similar to the behavior we expect from a natural attorney or even better than that. The AI will also learn from its failure or success. When a law student does not score very high in the examination, he or she learns how to improve the score in the next examination after analyzing the previous points of failure. The same aspect can also be applied for improving the performance of AI in developing legal reasoning.

"Digital sister-in-law" is one nomenclature given to AI. This is because generally AI is a robotic device that thinks like a human, interacts like a human, and can intermingle with people seamlessly and can use these characteristics to develop faultless legal reasoning. The digital sister-in-law can be better understood with the following example:

> Instead of going to watch a movie, you could ask your sister-in-law to provide a list of reads with storylines similar to that of the movie. The sister-in-law will provide not only a list of interesting reads but also a review of the movie that you wanted to see. The review of the movie will help you decide whether that movie is worth watching. Similarly, a lawyer or a judge who want to understand the nitty-gritty of a case can simply ask the sister-in-law a question and can have a meticulous list of answers to such questions. It is difficult to find such a model of AI. However, "agent technology" is software that works continuously, efficiently, and autonomously in a particular environmental situation and includes an agent that can make decisions in different difficult situations, where a natural person will generally be confused or

frustrated in trying to come up with a comprehensive decision. Agent technology is also adopted in online shopping schemes to assist consumers in search of relevant products. Agent technology can also help in finding relevant book materials in a digital library. At the same time, libraries can take the help of agent technology for connecting with other reference material of other universities and can also connect with the information literacy system and instructions. Agent technology can be a successful teacher in the undergraduate scheme and education system.

It is interesting to note here that agent technology was introduced by Roy Balleste, an academic law librarian and director of the law library at School of Law, St. Thomas University, with the help of a virtual library assistance called "Page." This agent technology provides many other facilities, such as distance education, cataloguing, circulation, and references, as mentioned by Balleste. There is a huge scope for the use of sophisticated AI in the agent technology in the academic law library, with possible future uses in, for example, changes in legal education, changes in understanding law and legal system, and changes in the dissemination of educational dimension of law branch. However, to make the system useful and successful, the users must be oriented to such technological advancement.

The use of agent technology in an academic law library requires many other factors to be considered, for example, the ever-changing legal environment, budgetary constraints, the nature of people accessing the library, and the working pattern of the library. It is true that AI and intelligent agents are exciting concepts. To improve the library services in terms of quality, it should be the duty of the head of the library to not only introduce the latest technology but also update the system so that the people who rely on the library are updated with the latest developments, even though a library is a place that is always underrated in terms of its value. It is also true that agent technology or AI should not and cannot replace the human librarians as they always add value to AI. Academic law librarians are multiskilled and highly educated and if associated with the agent technology, they can enhance the quality of the service of the library multifold. Services provided by law librarians, for example, teaching law students in a variety of settings, offering referral services, directing the users on how to use the library resources, and organizing research for faculty members and for their staffs as well, cannot be replicated by agent technology so easily. However, while patrons continue to interact face-to-face with reference librarians, many of their questions do not require a librarian's expertise. For example, for access to references as available in the academic law library, patrons can take the help of the agent technology. The sophisticated AI of the future integral to the agent technology can be a meaningful introduction to academic law libraries for reducing the burden on librarians, who can devote time to the more complex issues associated with running the academic law libraries. Library scholars, such as Rubin, Chen, and Thorimbert, have stated that at times, the librarian may not be available, and at that time, the client may approach artificial technologies, such as agent technology, for all information regarding references or other academic materials available in the library. It is like a nurse who performs the preliminary work before the doctor arrives on the scene. Accordingly, the conversational agent, rather than the librarian, can answer queries by the client that are repetitive and tedious in nature.

At the same time, there will be some clients who may not find it convenient to speak to the academic law librarian and would rather depend on computer software as a part of AI to find answers to their queries. Therefore, the academic law librarian may be involved in developing the technology for such future use and the regulatory technology behind this.

If people have the preconceived notion that a technology is not for them, the technology cannot provide a better ambience than a human, and the technology would be full of defects, the introduction of agent technology or higher AI will be futile. The present write-up exposes the benefits of agent technology and suggests more sophisticated AI in academic law libraries. It is now accepted that AI in academic law libraries provides quality service in many ways to students, faculty, local bar members, and litigants not represented by lawyers.

8.5 Do We Have Any International Regulation on Artificial Intelligence?

At present, there is no regulation to regulate AI. It is believed that Isaac Asimov's three laws of robotics continue to govern AI. The three fundamental laws of robotics are:

(1) A robot may not injure a human being or, through inaction, allow a human being to come to harm.
(2) A robot must obey the orders given to it by human beings, except where such orders would conflict with the first law.
(3) A robot must protect its own existence, as long as such protection does not conflict with the first or second laws.

These three laws were developed in 1950 by Asimov in his science fiction masterpiece entitled *I, Robot*.

The significance of these three laws is that they clearly specify that a robot having no such AI will not follow any of the three laws in its operation. Therefore, without AI available in the robot, these legal principles have no significance.

Historically speaking, arguments of cases were pleaded by the parties themselves. Subsequently, the concept was introduced that a person who is well versed in law should represent the parties before a judge for developing legal reasoning to arrive at a conclusive decision. It was in the seventh century when the practice of law converted to a real legal profession with people with great skills in developing legal reasoning. In the Middle Ages, this legal profession was almost on the verge of collapse but somehow survived, and modifications continue to this day.

It is also true that though the legal industry and law have been around for many years, they are now facing a transition because of the continuous intervention of technology in human life. The smartphone has become an integral part of human life, something no one would have thought remotely possible 10 years back. Similarly, the role of AI in developing legal argument and legal analyses for winning a case has garnered good recognition in the legal profession [11]. Nowadays, the dependency on technology is very high, with the Internet, AI, driverless cars, and drone deliveries just a few technologies worth mentioning [12]. At the same time, laws have been made to prevent the misuse of such technologies that could have a negative impact on the societies, for example, laws to prevent the use of the drone technology for carrying missiles and laws to prevent the misuse of the Internet. Basically, law and technology depend on each other.

A covenant on the regulation of the Internet globally was first established in the year 2001, when the Council of Europe Cybercrime Treaty for the first time developed the international platform to control the functioning of the Internet [8]. The United States did not pay any attention to the International Treaty on Cybercrime, and at the same time a lack of competency or legitimacy of the covenant put the international regulation in cold storage. Though international lawyers, the United Nations, and various human rights bodies demanded the establishment of international regulation of the Internet, raised a number of questions, and requested the states to take actions, ultimately the effort was futile. So, the Internet is becoming increasingly powerful day by day because of the lack of any consensus on international regulation [13].

The legal application of AI is another factor for consideration. For example, IBM's highly rated ROSS intelligence (developed by team ROSS), Watson has the potential to develop brief, concise, and appropriate answers for any natural language legal questions. There are software such as Rocket Lawyer, LegalZoom, and other legal service providers that operate online and provide benefits to their clients on many issues, for example, basic wills, divorce agreements, and contracts, without involving a natural lawyer at any stage of the process. Historically speaking, one had to use a lawyer for contesting any legal issues, including matters connected with family affairs. But now, computer software are replacing human lawyers in many parts of the legal profession. According to an analysis by Deloitte Insight, more than 1 lakh people will be replaced by AI machines and algorithms within next two decades in the legal profession. It has been found that the jobless sector would cover almost 39%. Under AI, there will be faster work in law profession, covering areas such as e-discovery, contract analyses, case prediction, and document automation, and the work will be cheaper. Reinforcement learning is one area where scientists involved in developing AI face a real challenge because the process involves analysis of past experiences, past errors, and past successes to reorganize and re-evaluate a case and develop more convincing and appropriate legal reasoning. The question is, how do you implement reinforcement learning? The interaction between the legal system and robotic software is of immense importance. However, when there is a need to fix responsibility, whether there was the developer's negligence or reasonable foreseeability of such harm would be taken into account. It is important to mention here that in *Jones v W+M Automation Inc.* (2006), a case from New York, the court did not find the defendant responsible because the defendant had taken due care as per the available regulations in force, though workers got injured because of the robotic gantry loading system. Under reinforcement learning, there is no liability on the part of humans and there is no foreseeability of causing harm. Accordingly, one cannot fix liability under traditional tort law. It is good that because of this unintentional development of law, the courts will have to adopt changes in the new technologies for appropriate decision and implementation of a perfect justice delivery system [14]. There is no concept of a sudden change in technology, especially in the era of the Internet, where e-mails are available and online research databases are also at your fingertips. For example, Lexis Nexis, Hein Online, and Westlaw are AI-based practices and are robust and efficient. These online legal databases are also changing their patterns regarding advanced search options, based on which the client can have multiple but appropriate answers in a very short period of time. A machine learning system is a unique feature because of the fact that the future challenges and future course of action can be dealt with efficiently under this scheme of reinforcement learning [15].

The scope of legal provision has changed drastically with the introduction of AI. On the one hand, the advanced technologies have helped to develop the legal profession professionally, and on the other hand, there has been negativity because of the introduction of such advanced technologies in the profession. Intellectual property (IP) law is a branch of law that has faced a tough time dealing with technological innovations. Technology is intruding into the day-to-day life of human beings and, therefore, requires proper checks and balances with the help of legal regulation. IP incentivizes the inventor toward unique inventions, and a lack of incentive provides less or no scope for generating interest in new inventions. Work relating to copy, transfer, and transformation has become faster, cheaper, and more private in nature, making it difficult to detect any infringement. Many structures have been created within information technology by the traditional enforcement machineries. David Levine, a professor of Elon University, stated that it is very difficult to translate and apply the IP law to intangible products moving through the Internet. At the same time, it will be difficult to fix the jurisdiction over the impersonation of IP rights because IP is internal in nature whereas the intangible products moving through the Internet are global in nature. Apart from this, the liberal people and the community who use the Internet for anything and everything protest against government agencies coming up with any stringent laws to block Internet operation. Because of such factors, comprehensive legislation to regulate the Internet world is not possible.

Since AI comes under the purview of a computer software program, its copyright is given to its creator. The Berne Convention, organized by the World Intellectual Property Organization (WIPO), prescribes protecting the copyrighted material identifiable as a qualifying work in a tangible form, for example, on paper, on a silicon chip, or on film. The convention further explains the expression "qualifying work," as, for example, a literary work, a music composition, a film, a computer program, a painting, or the expression of critical ideas. Accordingly, as per this convention, a computer software program is also copyrighted material. The Berne Convention, signed in the year 1886, is the oldest one and provides two basic principles, national treatment and automatic protection [16].

The WIPO copyright treaty, signed in the year 1996, is a crucial treaty related to computer software programs. According to it, computer software programs and databases both are protected under copyright law and any transmission of the work on a medium such as the Internet will come within the purview of the copyright protection provided to the original creator of such work [17]. There are a few other treaties established by the WIPO, for example, the Beijing Treaty on Audiovisual Performances, the Brussels Convention Relating to the Distribution of Programme-Carrying Signals Transmitted by Satellite, the Convention for the Protection of Producers of Phonograms Against Unauthorized Duplication of Their Phonograms (also known as the Geneva Phonograms Convention), and the Marrakesh VIP Treaty (formerly the Marrakesh Treaty to Facilitate Access to Published Works by Visually Impaired Persons and Persons with Print Disabilities) [18].

The future is waiting for legislators to implement a law for fixing responsibility against the developer if the AI machine causes harm to a human being and the environment. There is no denial of the fact that the future will be robust with these autonomous machines for various tasks, which cannot be completely defined right now [19].

8.6 Should Artificial Intelligence Be Taught to Law Students?

AI and law seminars have a pedagogical value mainly because the steps in the AI and law programs transform legal problems (input) into well-explained and convincing solutions (output). The present law schools across the globe may give an opportunity to law students to choose whether AI is the best option or human intelligence is the best option to provide a solution with a comprehensive explanation.

In the course of training law students on how to develop legal reasoning, they are informed about the important steps and while going through those steps, the students themselves identify the gaps that are to be filled in order to provide an appropriate explanation. The students have to identify whether the gaps are implicit, unexplained, and unjustified or opposed to these findings. It has been observed that when a student is developing legal reasoning against a hard question of law, the task of the student becomes more meritorious than when the student is developing the same for a soft question of law, where the legal reasoning and its development depend on the primary findings of law and are not too complicated. For a law student to be able to efficiently develop legal reasoning, it is better to have gaps that will encourage learning so the student of law has to consider the available relevant legal literature to fill the gaps, learning efficiently the mechanism of developing comprehensive legal reasoning in the process. In short, AI will help the student to develop a way to create his or her legal reasoning with the help of those gaps and by learning how to fill those gaps with the help of available legal principles and case laws.

The computational model involves AI and law research, and this program can help even that nontechnical law student who has no extra knowledge of computer programming. It is the algorithm that will describe the solution to any given legal problem while analyzing a vast quantity of legal material and from that, the law student will learn how to solve legal problems.

The law student can learn legal reasoning from the algorithm, which is an integral part of AI. As noted above, to learn how to successfully develop legal reasoning and to learn about the factors responsible for failure, it is important to study input/output examples for developing an automated legal reasoning program in AI [20]. The law student can best learn from failure because the system can show the student the gaps where relevant legal materials should have been mentioned for developing better legal reasoning. The stepwise development of argument and explanation in the form of legal reasoning can be understood under the thought experiment, and in this process a law student can learn how to develop a relevant question and how to develop a decision with the help of a comprehensive and convincing explanation.

To understand the computational modeling of knowledge of law, it is suggested that the law student know how to conduct a legal search, how to make a legal inference, and how to explain it in a better way. All these are a part of technical specifications and, therefore, require technical expertise on the subject matter. It is also true that a law student need not be a scientist to know these technical specifications but he or she does need expertise in how to develop legal reasoning based on legal research and legal inference.

To learn the technology associated with modern legal practice, more emphasis should be given to AI and law seminars, which have a pedagogical value and under which, the law student can learn efficiently in a conducive environment. In AI, there will be access to data mining, natural language, and processing of information on the use of case laws, which an attorney will know easily and will exploit the machine as per need. Therefore, a law student trained in this sector will also act like a professional attorney to identify the appropriate legal reasoning to develop a solution with an explanation. A law student will also have to be conversant with e-discovery tools for achieving the professional standard of care as demanded by the large corporate houses.

AI is a system that learns automatically from larger data sets, data mining, and natural language processing, which are quite complicated and can be beyond the imagination of a human lawyer. Therefore, attorneys or judges will have to efficiently use tools that will help them to interpret their outputs and bring convincing explanations to come to a decision. It is also important that unless a law student knows the techniques that work efficiently for providing legal reasoning, he or she does not rely on the AI system. The performance of the law student can be improved tremendously if such a student knows all the technical nitty-gritty and uses it to develop arguments beyond question. There are three ways to improve the efficiency of working with AI: (i) knowledge of the computational algorithm and its description, (ii) examples of when the algorithm achieves success, and (iii) the ways in which performance is calculated. Once conversant with these three parameters, the law student will definitely improve his or her performance in developing legal reasoning.

Law schools such as Chicago Kent, Pittsburgh, Northeastern, Stanford University, and Harvard University have been offering AI and law seminars at regular intervals for more than two decades. The International Conference on Artificial Intelligence and Law is a platform where students of law present papers regarding the viability of AI. The law seminar International Conference on Legal Knowledge and Information Systems is another platform where students present papers on the adoptability of AI and legal reasoning. The journal *Artificial Intelligence and Law* is also worth mentioning. The seminar papers address the issue of intervening steps, including input/output examples, of legal reasoning. The paper further deals with the human rationalization to be adopted in the case of identifiable gaps, based on which legal education can be imparted to identify how to fill the gaps by developing perfect reasoning. However, this system requires efficient knowledge of techniques dealing with AI.

It is always expected that AI and law seminars are a combination of students of law and students of computer science. This kind of program is mutually beneficial because the law student can provide legal input to the nonlaw student and the computer science student can provide technical insight to the law student on how to operate the AI system efficiently.

There is an advanced model of the AI and law practicum where the student of law will be required to learn the computational model of legal reasoning by constructing the same and applying the same for developing legal reasoning. AI and law practicum courses can be found in Stanford University, Chicago Kent, Pittsburgh, and Northeastern University, and the universities have been dealing with such courses for the last 10 years to develop software development tools basically for nonprogrammers. Therefore, an AI and law seminar can have clinical value after adding this law practicum, which is nothing but a "lab." This process of learning is a trial-and-error method because a student of law, whether having programming skills or not, can develop a legal expert system and by developing one, can learn how to fill the gap and even if the student fails, he or she comes to know what those gaps are and what legal materials are available for bridging the gaps.

A law student can develop a higher level of design even without having computer program skills. On the other hand, a computer program requires a refined designing process and, at the same time, must focus on the goals and subgoals while defining the steps for achieving the goals. Hence, it requires proper knowledge of computer programming. A student of law must think scientifically and systematically of a process for providing a solution to a legal problem and can devise a stepwise process for arriving at the solution for the problem to reach the goal or subgoals. With AI, a law student does not need to know every detail of the computer program or everything about law. All a law student should know is how to input the legal problem in an AI system and how to raise the questions based on which the AI will provide meticulous answers with explanations.

Therefore, it is clear that a law student does not require special skills to operate an AI system for developing legal reasoning because real programming in computer code starts only at the end of the process. The advanced AI system is built for the purpose of law students who have no sound knowledge of technology but can provide problems to this AI system and can generate specific solutions to such problems within a few seconds. It is possible to provide legal experts on technology, innovation, and law practice relating to law seminars organized by law schools for their students for understanding a system of AI by using a tool from Neota Logic, a system on which the law firms also depend for generating legal expert opinion for their clients.

It is interesting to note here that without having sound knowledge of computer programming, a law student can engage in designing a program and applying it for annotating legal text and this can be used for computation modeling of legal reasoning, which will establish a connection between the text and the legal significance. It is worth mentioning that the students at Hofstra law school learn about legal reasoning from empirical research work, which includes close study and annotation of evidential reasoning in legal context.

8.7 Story of IBM's Watson and Law Firms

With the extensive growth rate of advanced technologies, the scenario is not too far in the future that medical services will be completely under the control of AI and automation technology will take complete charge of the operation theater. Here is a simple situation: A patient goes to the hospital and instead of standing before a doctor stands before IBM Watson, a supercomputer in medical science. Watson collects each detail of the patient, from the date and place of birth to the environment under which the person was born, and with the help of a blood sample, identifies the person's genotype, frequently attacking diseases, and diseases on the parental side. Watson then analyzes the problem and provides an accurate prescription and medicines to the patient within a few seconds, where a human physician would have taken a week or more to carry out the entire exercise.

It is well-known fact that IBM developed Watson for taking part in the US trivia gameshow Jeopardy. However, a healthcare team is now believed to be using the same Watson in medical science for treating patients within a few seconds, which will be not only cheaper but also faster. It is also true that Watson is a computer program with a robotic feature that understands the natural language and connects the text with billions of pieces of material on medical science available in the cloud system and can access the personal data of the patient and a huge database of medicines having fewer side effects.

Watson can perform its innovative role in many areas, for example, social services, the healthcare industry, wearable technology, banking sector, data analytics, and even fantasy football. One can find out all these details by visiting Watson's Web site.

Watson achieved success in the area for which it was built, and now a lot of investment is being made in the system to take over services in many other areas for faster and cheaper relief. As per an IBM report, it took around five years to build Watson as a part of AI. The importance of Watson was realized in the trivia gameshow Jeopardy by the people and also computer scientists. The Watson business house was created in the year 2011 with around 107 Watson staffers.

Investors in Watson knew that it is the beginning of the entry of AI into human life, beginning with a mere gameshow and subsequent realization of its potential in the healthcare industry. For correctly judging and appropriately answering the questions, Watson has also created a valuable place in the legal profession. In the beginning, Watson was almost the size of a master bedroom but now it has come down to inches, doing work much faster in a shorter period of time and that too with great precision and efficiency.

Watson was used as a tool by the healthcare industry to diagnose the patient and prescribe medicine for accurate treatment. Once, when a researcher raised a query, Watson replied with "bullshit," a funny episode indeed! Watson used this expression because the dictionary used as the input material had a few defects in vocabulary. Later, the defective dictionary was removed from the database of Watson. It is also important that Watson display accurate answers in a short period of time.

Watson involves a trial-and-error method. When Watson was wrong in diagnosing a patient and prescribed the wrong medicine, the medical specialist corrected it by providing the correct analysis in its computer program. So making Watson learn again and again is the doctrine, in a manner similar to "wash, rinse, repeat."

Watson has been used by the financial market as well. For example, Citibank was the first financial bank that used Watson to improve the customer services provided by bank employees, to identify the premium clients who would pay their loans in advance, and also to figure out clients with fraudulent intentions [21].

8.8 ROSS: A New Venture of AI

Watson is an example of AI that satisfactorily answers questions raised before it. It is a system that tries to understand the natural language of the question, analyzes each and every text of the question with the available material within itself and also in the cloud system, and then provides a precise answer to the question.

In a Watson University competition sponsored by IBM, a group of students of the University of Toronto developed a new legal application of the Watson platform, called "ROSS, the super intelligent attorney." The students won the second prize. IBM referred to this higher-level machine for the legal profession market as "the son of Watson."

It was IBM that allowed the ROSS team to have continuous access to the Watson platform, which is basically a cloud platform. A large number of public legal documents were uploaded to the ROSS database with proper identification of the subject matter, expert opinion, and legislations to answer any sort of queries on law. However, it should be mentioned that Watson is not powerful because it gives accurate answers; rather it is powerful because it has an automatic ability to learn from its failures and also from successes, including feedback given to it. Therefore, the more you use Watson, the more powerful you make it [22].

The working of ROSS depends on the platform that Watson provides in terms of understanding the natural language and also accessing the cognitive computing platform. The creators of ROSS claim that it is funded by Dentons, a firm with more than 6000 lawyers. ROSS will become a partner of every law firm in the near future because of its ability to prepare affidavits, case briefs, affidavits in reply, and arguments versus counterarguments.

ROSS is in the process of learning US bankruptcy law, which is a special subject for the attorneys in the United States; all the cases related with this; and the interpretation of the statute. ROSS is a part and parcel of many high-profile law firms, and the former students associated with the development of ROSS are now entrepreneurs behind its intelligence.

It was reported in a newspaper that Dentons never discloses the investment behind ROSS Intelligence's Nextlaw Labs, a project with the objective of developing a new computer program for the legal industry. Dentons, in association with IBM, teams with legal startups, like ROSS Intelligence, and also works within Nextlaw Labs for advancement of such technology to be utilized by the legal industry in the future. Nextlaw Labs use IBM's cloud computing resources for improving the advanced technology to be utilized in the legal industry.

AI and law is a subfield of AI, and in this area many technologies are available, most importantly, IBM's Watson debate. This debate is one of the unique features of IBM's famous Watson computer. If Watson is asked any question on a legal topic, the computer programs available within it automatically and autonomously scan its huge database of knowledge as identified in natural language and texts relevant for answering the question and after understanding the data, it generates the strongest and most favorable argument, with the explanation to provide the correct and appropriate answer to the question raised. This system can help lawyers with quick access to case notes, precedence, and relevant legal materials across the globe, including any justification made by Watson.

It is interesting to note here that "Siri for the law" was also developed in IBM's Watson labs. Like we have the Google search engine for searching any material available on the Web, ROSS is a legal research tool that enables lawyers to get answers in legal language from the billions of legal documents available on the Web. It has been witnessed that many a time, Watson will provide the answer in general English rather than in legal terms. The credibility of ROSS is at an unprecedented level because of the fact that it can respond to legal citations and suggest relevant articles for further reading, cases, and citations of the cases by judges. A cognitive computing system is a system under which a computer program learns from its past. This practice enhances the accuracy level of ROSS in any future course of action. It reduces the time the lawyer has to spend on a particular case and provides sufficient legal documents to argue the case before the judges.

An Israeli startup called ModusP has established another level of search engine, one that uses a highly sophisticated algorithm relying on AI. This search engine helps jurists, lawyers, law students, and even the layperson to search for any relevant legal material within a few minutes, providing plenty of supporting documents, which not only enhances knowledge but also provides answers efficiently.

Lex Machina is an IP research company that helps the company to presume, find, manage, and win patents and other IP lawsuits by making comparative analyses with an IP database bank. This company helps customers to come to valuable conclusions. This technology mainly collects information on cases already decided and statutory provisions interpreted by the courts and analyzes the questions to provide appropriate answers. Under this technology, a lawyer can collect appropriate information on specific judges and the history of the client.

A cloud-based platform has been developed by eBay and PayPal by the name of Modria, which works as an online dispute resolution platform, where the business houses can receive a quick resolution to their dispute, without the company interests being affected. It has information on billions of cases and statutes on the basis of which the system provides a solution to any legal problem. An AI named "Premonition" provides data on how many cases have been won by a particular lawyer and before which judges.

Proper counselling on a contract can be done by "Beagle," an AI that can quickly highlight the most important clauses of the law of contract and can provide collaboration in real time. This also enables easy negotiation by the lawyer and can collect quick feedback from its client for further improvement. An AI named "Legal Robot" works on the law of contract. This platform enables lawyers to understand a legal document, a legal agreement, relevant cases, and relevant judgments and to develop arguments with the help of supporting legal documents.

8.9 Law Practicum and Artificial Intelligence

Law practicum is nothing but a lab system under which the students of law are trained to use computational models of legal reasoning. Law practicum courses are offered in a number of universities, for example, Northeastern University, Chicago Kent, Stanford University, and Pittsburgh. Under this scheme, a student of law with no knowledge of computer programming can also do well and develop comprehensive legal reasoning with the help of AI.

Law practicum helps students to learn the law from textbooks and helps them to apply this knowledge practically, which is nothing but providing legal education with the help of AI [23]. Under this program, a student can device a new computer program, with the help of which he or she can develop legal expertise on the subject given. There are factors under law practicum on which the law student should focus, such as participating in diagramming arguments and annotating legal documents, experimenting with applying predictive coding, participating in Trec legal track, reasoning with cases, and dealing with uncertainty, so that a computational model of the legal argument can be developed by the AI.

8.10 Methods of Computation for Developing Reasoning with Legal Rules and Cases

Edwina Rissland made a survey regarding the introduction of AI to law students, including application of the legal domain and how to use the computational model of legal reasoning. Donald Waterman's expert system encourages heuristic rules on product liability, and the second model of Anne Gardner faces questions from the law of contract law. Therefore, the syllabus relating to the introduction to computational model of legal reasoning includes these two examples for new law students to understand the implications of the application of AI.

The above-mentioned computational model and the AI machinery draw a very specific legal conclusion. For example, Rissland exposes the examples given by Waterman and Gardner and also focuses on the case-based model of legal reasoning.

Waterman devised and developed three kinds of rules: those for legal authorities, those for litigators, and those for insurance claim adjusters. Many other factors, such as state liability, comparative negligence, product liability, and rules guiding settlement computations, will come under this segment [24].

Gardner's program is more on the law of contract in relation to offer and acceptance. It comes up with conclusions—for example, where there are no relevant legal relations, an offer is pending, a contract exists, or a contract plus a pending proposal to modify the contract exists—that are analyzed with reasoning by the AI.

Marek Sargot developed the logical models of statutory reasoning that analyze legal questions involving British citizenship. Under this system, determining citizenship becomes a computable item. Under this system, any citizen can ask questions on citizenship and would be given satisfactory answers to those queries.

Computation of a large database of statutory rules becomes difficult when the statutory rules are amended frequently. To avoid this practical problem, a project was developed by Trevor Bench-Capon and Frans Coenen where the statutory code was isomorphic in nature.

Logical rules may be another difficulty with statutory texts. Because natural language and text will be different from the statutory code, the interpretation by the AI may create difficulty and logical ambiguity. Another difficulty is that once the AI machine provides a conclusion from the given legal references, it cannot be dismantled or taken back. Therefore, due care should be taken to build the computer program as a part of AI so that there is flexibility in devising legal reasoning and conclusion.

Somewhere, there exists a model for predicting legal outcomes and case-based models of legal reasoning based on pure rule-based legal reasoning. There will be flexibility in developing legal concepts and providing conclusions. Therefore, the terms of the directions on the stereotypical fact either strengthen or weaken a claim. There is also a computer program that demands the involvement of semantic networks that will be representing facts that generally the judges wish explained in their judgment. This is known as the GREBE model.

This program develops arguments for and against the proposed conclusion on inference, which is different from the other models. This is quite similar to HYPO's three-ply arguments.

The development of computational models empirically is based on case-based models and is quite similar to the Turing test of GREBE. A human expert was appointed to create the law students' paper and also was not informed that there is a computer program that will also generate the paper like a law student. When the grader was given the student paper for gradation, he could do that very well but was not aware of the fact that the concerned student paper had been basically developed by the computer program. This program was once deployed by Karl Branting.

A factor-based approach influenced by case-based models, CATO and IBP, can make a prediction on a legal case with the possible outcome depending on the legal reasoning developed from this database and available in the argument. This model became an example of the predictive accuracy of AI and law and can be practiced in Canadian tax cases involving capital gains issues. CATO and IBP can not only predict the outcome but also develop the argument by which the outcome can be reversed.

GREBE is a program integrated on rule-based reasoning—indexing semantic networks of critical facts in past cases and case-based reasoning—and it works very well in worker's compensation law.

CABARET is another level of program combined with logical representations of an IRS provision with the further dimension of understanding open-textured statutory terms. This program is a combination of rule-based and case-based analyses and is applicable in laws connected with a taxpayer's position.

Donald H. Berman and Carol Hafner, considered to be influential critiques of the case-based computational model, stated that this model is deficient in terms of the teleological component. This means that the purpose of the computer program is to help in reasoning related to legal statutes and rules. This criticism ultimately forced the development of a new model with greater success, like the model of legal reasoning with cases and precedence developed by Bench-Capon and Sartor. Another task to be performed by AI is to identify how case decisions extensively define the principles in professional ethics. It works more on semantic networks of critical facts in understanding ethics. Bruce M. McLaren developed his SIROCCO program, where cases and ethical rules would be combined to arrive at an inference [25].

8.11 Scheme of Argument and Legal Reasoning

AI and law research works on the computational framework for nonmonotonic, defeasible legal reasoning and recasts the logical deductive, case-based, and value-based approach. The physical inference or nonmonotonic models can take back previously concluded outcomes once the new information is available to the system.

It is important to note here the assessment of legal claims with the argument schema developed by Katie Atkinson and Trevor Bench-Capon and by Thomas F. Jordan and Douglas Walton. Evidentiary arguments are one of the important features of this system that can be used in standards of proof in any legal claim.

For a case comparison with value judgments based on intermediate legal concepts along with analyzing hypothetical cases while referring to US Supreme Court oral arguments, a new mechanism has been developed by Matthias Grabmair, also known as the argument schema.

For improving the legal argument, the program argument schema is very effective. McCarthy developed a flexible conceptual schema for demonstrating empirically intermediate legal concepts. A program schema is flexible in terms of its own learning by its past activity and also relearns any modification in the existing statutory norms.

A domain concept with explicit and systematic specifications and used in legal rules and factual situations helps to make the computational model flexible. Ontologies also make the computational model flexible, with a case representation to make the system work in contexts. Legal ontology is a process where the development of an analogy, teleological, and hypothetical legal reasoning and the use of argument schema are worth mentioning.

8.12 Reasoning with Open-Textured Texts

AI can develop logical and legal reasoning based on full-text legal information retrieval (IR). Probabilistic models of retrieval followed by Bayesian network methods are used for evaluating legal IR for developing legal reasoning based on available factual situations.

There will be advantages as well as disadvantages of full-text legal IR and AI and law models once legal IR and AI have been combined. SPIRE is a program of AI where case-based models are used to automatically find relevant legal passages. There is a network—SCALIR network—where legal cases are linked via shared terms or citations. This network also provides sufficient material for developing legal reasoning on the basis of case database analysis.

The extraction of legal information and processing of text is possible when legal reasoning and legal texts are connected through a computational model. The classification of textually described cases can be found on the Internet under the project titled SMILE+IBP.

Legal research can presume that the legal text available in any case database has a fairly homogeneous structure from which an appropriate inference can be drawn, and that is the function of AI and law, to work on legal information extraction and text processing. Extraction and text processing may not apply to electronic discovery because the parties to a lawsuit should not only analyze the given situation but also produce a number of relevant legal documents for or against the allegation. The evidence relevant to the lawyer's hypothesis about claims and documents is one of the works of AI and law tools for electronic discovery [26]. The IR tool will definitely take a lead role when there are high volumes of documents available for assumption and determination of the legal reasoning out of domain information. AI tools are always important for a company in discovering documents that will be unique for winning the case. Similarly, AI tools may also play a significant role in filtering documents to get relevant legal documents.

8.13 Reasoning with Cases, Hypothetical Situations, and Precedence Citing

Unless liability is fixed on the developer of the AI for wrongly suggesting an inference or advising the client, which ultimately becomes troublesome for the client, the system of AI will not be foolproof. It is also an accepted fact that in this era of technological advancement, the legal system has to make use of AI services. Therefore, there is the possibility that the AI will fall into the scope of the court system, where the liability from the tort law can be invoked if the AI system fails. The court may be the final authority for giving full permission for the use of AI in legal reasoning and legal development because the liability regime is uncertain and how to fix responsibility in this field is still being explored. It is also not easy for legislators to immediately devise regulation and control the AI system; rather it will be done on a case-by-case basis while continuously improving the standard of AI regulation.

The court system will be effective in using AI not only for argument but also for the delivery of justice by the judges. But the truth is that there is a serious weakness in the judicial system in that the system itself is neither aware of the importance of AI nor technologically trained in the system to operate it. The advanced technological system that will help develop legal reasoning must be taught in the law school to the law student so that this awareness about the application of AI reaches the judicial system when the student is selected in judicial services or practices as a lawyer in the court of law. In this way, AI will be integrated in the judicial system after a few years but there is no immediate relief in sight involving the use of AI to decide cases. Judges have not yet understood how AI is going to change the way we do business, as stated by Chief Justice Roberts, Chief Justice of the United States. The mindset of the legal system must be changed—which generally accepts the fact that there will be delay in deciding cases—that after the introduction of AI, cases will be discussed at a faster rate and decided accordingly. The lawyer's fees will be reduced because lawyers charge fees per appearance and AI will reduce the scope of such appearance, which can be grounds for agitation by lawyers. At the same time, the judiciary is remedial in nature, that is, it either corrects the person who is criminally charged or provides compensation for any civil wrong. Therefore, AI cannot reduce the time of litigation that is lengthy because of procedural difficulty; rather the argument supporting a decision may come faster but AI cannot help reduce any delay because of the procedural law. Another limitation is the narrow approach of judiciary over the application of AI, which will definitely reduce the application of the AI system. This cannot be the ground for a superiority complex; rather it should be looked into as a feasible option both for the court system as well as for the clients who have filed cases before the court having understood that the cases will be decided at the earliest.

8.14 MOOCs: Is It an Example of an Intelligent Tutorial System Coupled with Ethics?

Massive open online courses (MOOCs) are another example of the application of AI, where the students get an interactive online platform related to the respective educational Web site. The system is so advanced that students can post questions on this platform and the computer program will provide perfect answers to those queries. If the educational institution adopts a policy like the Watson platform, then with the cognitive model the computer programs available in MOOCs can also learn from past experiences—failure and success.

In the system of asynchronous learning, a massive number of students can enroll for online courses, appear for the examination online, and appear for viva exams, all this simultaneously. It is important to mention the year 2012 here because this was the year the term "MOOCs" was coined and the concept practiced by some educational institutions and a few students. However, because of the success of MOOCs, professors from the United States started offering various courses under MOOCs. Courses for the undergraduate scheme of MOOCs became famous, such as the environmental law course by the University of North Carolina professor Don Hornstein, the constitutional law course by Yale University professor Akhil Amar, the international criminal law course by Case Western reserve Michael Scharf, the copyright law course by Harvard University professor William Fisher, and the course on law and entrepreneurship by Northwestern University professors Esther Barron and Steve Reed. MOOCs got hugely successful feedback from the law schools that offer online courses under this scheme, as noticed by Professor Philip Schrag. AI application in MOOCs will take over the traditional teaching system of law schools, and a few law schools with a traditional setup will survive because of the students not open to accepting online courses. It is also true that through MOOCs most of the students will do introductory courses on law rather than detailed study as available under the traditional lecture-based model. This was evidenced by researchers from MIT, Harvard, and Tsinghua University. With further advancement in AI and law, there is a possibility of striking a golden balance between the traditional lecture method of teaching and imparting of education through an online platform.

8.15 What Future Do We See of Legal Reasoning Associated with Artificial Intelligence?

The initial phase of AI has been developed by researchers over a period of time. The development of AI to the extent that it can interact with the Internet system as an agent as per the requirement of the client started in the mid-1990s, when people thought to experiment with AI, which can play a bigger role in daily life. AI has the ability to understand the natural language for further analysis and because of the continuous research in the area, analysis of natural language became successful, which is evidenced by Apple iPhone's Siri, Google voice, Microsoft dictation, and Dragon NaturallySpeaking software. It is important to mention here that so far, the Dragon NaturallySpeaking software is the best example of personal assistance in relation to the computer. Writers in the medical profession, in the legal profession, or for celluloid can definitely rely on Dragon NaturallySpeaking software. The software has been further improved since its development, and the speaker can improve the dictation quality with the help of the software. Dragon NaturallySpeaking software is an example of AI where speech can be converted into text and speech can be converted into action. Self-driving cars are another example of AI, as is Google, which developed Google Maps, a software program that acts as a personal assistance that with its own intelligent technology can advise the person on the best route to take. A deep learning research project has been initiated by Google Brain. Similarly, a computer program that gathers common sense while using the image database is an example of advanced AI as developed by Carnegie Mellon's NEIL project. It was in the news that Google has invested US$650 million for purchasing mind technologies, with a focus on higher research on AI. At the same time, a robotics company in the name of Boston dynamics has also been purchased by Google without disclosing the amount invested. It is clear that Google is a company that is serious about improving the functionality of AI, which can be used for mankind at a more convincing level.

AI thinking like a human, for example, a cognitive model, is a major focus of further research on AI, and experts are counting the days when such AI will achieve its full functionality, like humans. Accordingly, researchers are busy and investors are pouring money into this project with the hope to develop a state-of-the-art algorithm that can bring an AI at par with human intelligence in the near future. Irrespective of what scholars say, optimists and skeptics alike are of the view that achieving a fully functional AI may take 80–400 years. However, with the rapid growth of the Internet and extraordinary achievement in technological advancement, researchers are hoping that soon fully functional AI will be available to serve mankind.

8.16 Conclusion

From the discussion above, it can be inferred that the daily life of human beings depends a lot on the facilities provided by AI. The calculator is an example of low-level AI, whereas diagnosing a patient and providing a prescription along with medicines is an example of a higher-level AI.

1950 onward, Turing, McCarthy, and Balleste are a few important names associated with the development of AI.

The jurisprudence of AI is the phase of acceptability in human society, when such AI understands the legal question in the form of natural language, examines from its database what could be the possible connecting words, and provides a meticulous answer with many explanations and justifications.

Legal reasoning developed by AI is one of the experimental phases of understanding how a computer program can replace the lawyer, because AI can develop the argument and answer the questions with supporting documents within a few seconds, which a human lawyer would take weeks or months to do. The judges in the legal system depend more on a critical analysis of the given factual situation and prefer the argument of that lawyer who provides sufficient explanation while answering queries raised by judges or opponent lawyers.

There are many developments in the field of AI, such as the comprehension test, the Winograd schema challenge, the logic theorist system, and the Lovelace 2.0 test, which are examples of nonhuman computational abilities [27].

The Turing test was one of the fundamental and earlier tests to find out the physical applications of AI for humans. Computational modeling and the cognitive model, both were researched under this scheme, and it was found that the cognitive model of AI can think like a human and work like a human.

IBM's Watson was first noticed when this computer program defeated the then champion of chess Gary Kasparov, and that computer application program was called "Deep Blue." However, Watson became popular and caught the attention of researchers interested in carrying out research on AI when Watson won the first prize in the US trivia gameshow Jeopardy. Experiments in the field of medicine were carried out on Watson, with Watson providing advice to patients and also treating patients while analyzing their medical histories and providing appropriate medicines by prescription [28].

Watson was also taken into consideration for developing legal reasoning, which was considered to be the jurisprudence of AI in the field of law. The platform of Watson was so efficient that many legal documents were uploaded to its system and on the basis of those legal documents, any question in natural language on law could be diagnosed by Watson and connected with billions of pieces of relevant texts available in its database to find the correct answer with comprehensive explanation and justification.

A more advanced AI technology, ROSS, was introduced by IBM, which was working on the Watson platform to provide answers meticulously. ROSS became more popular than Watson for providing comprehensive and meticulous answers to legal questions posted before it. ROSS was in the position to develop arguments with proper justification within a few minutes that a human lawyer would have taken months to develop.

With the presence of a high level of AI, there is a possibility of law firms replacing many human lawyers and there is also the possibility of judges showing the tendency to rely more on AI than human lawyers for faster access to relevant legal materials, decisions, and explanations.

Copyright laws are enacted by national-level legislatures. Therefore, there is no international regulation dealing with copyright issues related to computer programs. As we consider AI a computer program, it would be wise to consider that only the Berne Convention, organized by the WIPO, may be of some help to understand the international scope of copyright protection of the computer software programs related to AI.

However, presently, it cannot be claimed that AI has reached its full potential or that it has intelligence at par with human beings. Researchers are of the view that soon the society will encounter AI in its fully grown condition and AI will be used to serve mankind keeping in mind the three laws of robotics as developed by Asimov.

References

1. Harris, J. (2019). Reading the minds of those who never lived. Enhanced beings: the social and ethical challenges posed by super intelligent AI and reasonably intelligent humans. *Cambridge Q. Healthcare Ethics*, **28**(4):585–591, doi:10.1017/S0963180119000525.

2. Lawrence, D., Palacios-González, C., Harris, J. (2016). Artificial intelligence: the shylock syndrome. *Cambridge Q. Healthcare Ethics*, **25**(2):250–261, doi:10.1017/S0963180115000559.

3. Baker, J. (2007). Intelligence. In *In the Common Defense: National Security Law for Perilous Times*, Cambridge University Press, Cambridge, pp. 126–175, doi:10.1017/CBO9780511509759.008.

4. DeGlopper, D. (1992). *The Art of Computer Systems Performance Analysis: Techniques for Experimental Design, Measurement, Simulation and Modeling* by R. Jain, John Wiley & Sons, New York, 1991, p. 720; *Int. J. Legal Inf.*, **20**(1):63–64, doi:10.1017/S0731126500010933.

5. Goldberg, M., Carson, D. O. (1991). Copyright protection for artificial intelligence systems. *J. Copyright Soc. U.S.A.*, **39**(1):57–75.

6. Ravitch, F. (2010). No intelligence allowed or no intelligible science? In *Marketing Intelligent Design: Law and the Creationist Agenda*, Cambridge University Press, Cambridge, pp. 157–200, doi:10.1017/CBO9780511761164.007.

7. Tecuci, G., Schum, D., Marcu, D., Boicu, M. (2016). *Intelligence Analysis as Discovery of Evidence, Hypotheses, and Arguments: Connecting the Dots*, Cambridge University Press, Cambridge, doi:10.1017/CBO9781316388488.

8. Buiten, M. (2019). Towards intelligent regulation of artificial intelligence. *Eur. J. Risk Regul.*, **10**(1):41–59, doi:10.1017/err.2019.8.

9. Ashley, K. (2017). *Artificial Intelligence and Legal Analytics: New Tools for Law Practice in the Digital Age*. Cambridge University Press, Cambridge, doi:10.1017/9781316761380.

10. Frydrych, D. (2017). Rights modelling. *Can. J. Law Jurisprud.*, **30**(1):125–157, doi:10.1017/cjlj.2017.6.

11. Waters, D. (1957). Chief Justice John Marshall: a reappraisal, by W. M. Jones, [Cornell U. P. (Oxford U. P.), 1956, 24s]. *Int. Comp. Law Q.*, **6**(3):576–579, doi:10.1093/iclqaj/6.3.576.

12. Cristina, G. M. (2019). Data protection and self-driving cars: the consent to the processing of personal data in compliance with GDPR, *Commun. Law*, **24**(1):15–23.

13. Cock Buning, M. (2016). autonomous intelligent systems as creative agents under the EU framework for intellectual property. *Eur. J. Risk Regul.*, **7**(2):310–322, doi:10.1017/S1867299X00005730.

14. Fidler, D. (2018). Synthetic brain technologies: beyond artificial intelligence. In *New Technologies and the Law in War and Peace*, Boothby, W., ed., Cambridge University Press, Cambridge, pp. 359–391, doi: 10.1017/9781108609388.014.

15. Bannerman, S. (2016). Institutional aspects of international copyright at WIPO. In *International Copyright and Access to Knowledge*, Cambridge Intellectual Property and Information Law, Cambridge University Press, Cambridge, pp. 201–219, doi:10.1017/CBO9781139149686.014.

16. Hristov, K. (2017). Artificial intelligence and the copyright dilemma. *IDEA: The Journal of the Franklin Pierce Center for Intellectual Property*, **57**(3):431–454.

17. Štrba, S. (2018). The Marrakesh treaty, public–private partnerships, and access to copyrighted works by visually impaired persons. In *The Cambridge Handbook of Public-Private Partnerships, Intellectual Property Governance, and Sustainable Development* (Cambridge Law Handbooks), Chon, M., Roffe, P., Abdel-Latif, A., eds., Cambridge University Press, Cambridge, pp. 176–198, doi:10.1017/9781316809587.012.

18. Palace, V. M. (2019). What if artificial intelligence wrote this: artificial intelligence and copyright law. *Florida Law Rev.*, **71**(1):217-ii.

19. Harris, J. (2018). Who owns my autonomous vehicle? ethics and responsibility in artificial and human intelligence. *Cambridge Q. Healthcare Ethics*, **27**(4):599–609, doi:10.1017/S0963180118000038.

20. Bénédict, W. (2018). Excursus: artificial intelligence. In *Responsibility, Restoration and Fault*, Intersentia, pp. 145–150, doi:10.1017/9781780686202.010.

21. Tee, L. (2001). Artificial intentions. *Cambridge Law J.*, **60**(2):231–264, doi:10.1017/S0008197301680620.

22. Gärditz, K. (2017). Legal restraints on the extraterritorial activities of Germany's intelligence services. In *Privacy and Power: A Transatlantic Dialogue in the Shadow of the NSA-Affair*, Miller, R., ed., Cambridge University Press, Cambridge, pp. 401–434. doi:10.1017/CBO9781316658888.016.

23. Weinreb, L. (2005). Analogical reasoning, legal education, and the law. In *Legal Reason: The Use of Analogy in Legal Argument*, Cambridge University Press, Cambridge, pp. 123–162, doi:10.1017/CBO9780511810053.006.

24. Samuel, G. (1991). The challenge of artificial intelligence: can Roman law help us discover whether law is a system of rules? *Legal Stud.,* **11**(1):24–46, doi:10.1111/j.1748-121X.1991.tb00621.x.

25. Gerdes, A. (2018). An inclusive ethical design perspective for a flourishing future with artificial intelligent systems. *Eur. J. Risk Regul.,* **9**(4):677–689, doi:10.1017/err.2018.62.

26. Walton, D. (2007). *Witness Testimony Evidence: Argumentation and the Law,* Cambridge University Press, Cambridge, doi:10.1017/CBO9780511619533.

27. Ayton, P., Billingham, P. (2007). Searching for the answer 2.0. *Legal Inf. Manage.,* **7**(4):258–262, doi:10.1017/S1472669607002071.

28. Heijmans, A. (1974). Artificial Islands and the law of nations. *Neth. Int. Law Rev.,* **21**(2):139–161, doi:10.1017/S0165070X00019793.

Chapter 9

Legal Ethical and Policy Implications of Artificial Intelligence

Subir Kumar Roy

Department of Law, P.G. Faculty Council of Arts & Science, Bankura University,
West Bengal, India
dr.roysubir@gmail.com

This article intends to discuss at length the issues and challenges before AI and the consequences of AL for this planet. It also intends to critically discuss the existing regulatory approaches to AI; the need of global governance; and AI and humanity under the spectrum of human rights mechanism, along with the accompanying ethics. The approach of the article is primarily doctrinal, based on review of literature, laws, and policies (though they are scanty, scattered, and almost negligible) and advocates in favor of global collaborative measures to regulate AI.

9.1 Introduction

Human beings are empowered with creativity. The human intellect and mind always quests for something new, and today's develop-ment, what we are witnessing, is the consequence of this creativity.

Artificial Intelligence and the Fourth Industrial Revolution
Edited by Utpal Chakraborty, Amit Banerjee, Jayanta Kumar Saha, Niloy Sarkar, and
Chinmay Chakraborty
Copyright © 2022 Jenny Stanford Publishing Pte. Ltd.
ISBN 978-981-4800-79-2 (Paperback), 978-1-003-15974-2 (eBook)
www.jennystanford.com

Humans depend on their power of creativity to create machines in their own image and bestowed with artificial intelligence (AI) and also to triumph over diseases and death by exploring the concept of artificial life (AL). At the same time, there is apprehension that these techniques may turn against their own survival, and for this reason humans tend to adhere to ethical and legal matters related to the consequences of such development.[1] AI is on a threshold and expected to lead human civilization with a new orientation that may change the traditional concept of life and working habits at a breakneck speed. Humanity is looking at AI with curiosity and doubt, fear and favor, anticipation and trepidation—in confusion and dilemma about whether to welcome it or to discard it. However, recent progress in AI has been splendid and exponential. We find the potential impact of AI on and its use in different fields, like machine translation, healthcare, aircraft operation, finance, defense, education, criminal justice system, space and aeronautics administration, and technology.

9.2 Genesis and Concept of AI

As far as the genesis of AI is concerned, we found that John McCarthy, assistant professor of mathematics at Dartmouth College had coined the term "artificial intelligence" while writing a proposal for the pivotal summer conference held at Dartmouth College in 1956.[2] This conference was considered as the first major concerted approach to pave the path for AI, where McCarthy, Marvin Minsky, Allen Newell, Herbert Simon, and others exchanged their views on the above.[3] Simon, who is regarded as one of the pioneers in the evolution of AI, had commented in 1957, "It is not my aim to surprise or shock you. . . . But the simplest way I can summarize is to say that there

---

[1] D'Alarcao, H. (1984). Artificial intelligence: myths and realities. *Bridgewater Rev.*, 3(1):13–16, available at http://vc.bridgew.edu/br_rev/vol3/iss1/7, accessed on May 18, 2019.

[2] Goldberg, S. (1994). Artificial intelligence and the essence of humanity, p-152 in the book entitled *Culture Clash*, NYU Press, available at https://www.jstor.org/stable/j.ctt9qfqbm.12, accessed on May 17, 2019.

[3] Ibid.

are now in the world machines that think, that learn and that create. Moreover, their ability to do these things is going to increase rapidly until—in a visible future—the range of problems they can handle will be co-extensive with the range to which the human mind has been applied."[4] Minsky had opined about AI that it is a science of making machines do things that would require intelligence if done by humans.[5]

Modern research has brought sea changes in the concept of AI. Computers are no longer only number crunchers; rather they can mimic human behavior in almost all areas. AI in the modern context has been defined as "Technologies with the ability to perform tasks that would otherwise require human intelligence, such as visual perception, speech recognition, and language translation."[6] It refers to a machine vested with broad cognitive faculties so that it can think and do work on its own convincingly, even surpassing human intellect in such a manner that it would essentially be indistinguishable from the activity of a human being.[7] So AI signifies the empowering of a machine with such intellect whereby it is impossible to distinguish the intellectual power of the machine from human abilities. The aim of the creation of AI is always to outperform the potentiality and capacities of the human being so that a complex situation in any field can be resolved in a timely manner, accurately, and in an inexpensive and articulate way. But the question is, when will humanity be able to develop "artificial human-level intelligence" and whether that intelligence, using the abundant

---

[4] Dreyfus, H. L., Artificial intelligence in the annals of the American Academy of Political and Social Science, Information Revolution, March 1974, p-22, quoted from Simon, H. A., Newell, A. (1958). Heuristic problem solving: the next advance in operations research. *Oper. Res.*, 6(1):1–10, available at https://www.jstor.org/stable/1040396, accessed on Sept 15, 2019.

[5] David Bolter, J. Artificial intelligence, *Daedalus*, Vol. 113, No. 3, Anticipations (Summer, 1984), p-1 quoted from 1Semantic Information Processing, by Minsky, M., MIT Press, Cambridge, 1968, available at https://www.jstor.org/stable/20024925, accessed on Sept 15, 2019.

[6] Report of the House of Lords, Select Committee on Artificial Intelligence, of session 2017–19, p-14, quoted originally from Department for Business, Energy and Industrial Strategy, *Industrial Strategy: Building a Britain fit for the future* (November 2017), p. 37 available at https://publications.parliament.uk/pa/ld201719/ldselect/ldai/100/100.pdf, accessed on Oct 15, 2019.

[7] Ibid. p-15.

available computational resources, will quickly bootstrap itself into superintelligence?[8] AI has been considered as one of the most adventurous and ambitious scientific and engineering endeavors of all time.[9] In modern times, research in AI aims at simulating the human brain to solve complex problems with logic and reasoning, striving toward conceptualizing the concept of AL. Researchers say that the situation demands an understanding of the mind from a new orientation and enabling AIs to learn and apply intelligence in different fields to obtain benefits from it.[10]

AI is already widely used but. At the same time, it has far-reaching social, economic, and ethical implications. It involves serious legal and ethical issues. The exponents apprehend that in the near future, AI may exceed human intelligence and become uncontrollable. Uncontrolled AI is presumed by many to be even more dangerous than nuclear weapons and may expedite a third world war. This situation may ravish the whole humanity and therefore requires addressing the issues of AI within the framework of the International Human Rights Mechanism. Besides this, the development of algorithms and uses of different analytical tools to deal with the huge amount of data may seriously jeopardize the right to privacy, security, safety, pluralism, democracy, freedom, and justice delivery system. Automated systems with AI will lead to a large chunk of people being laid off from their jobs and will reduce the number of workplaces. However, AI will increase productivity and may create job opportunities in other sectors. AI is expected to bring radical transformation in the health system, which may change the whole demographic policy. Revolutionary changes may take place in security systems, administration, and judicial approaches. Supercomputers may turn into dictators because robots cannot acquire human feelings. To prevent such a situation, "global

---

[8] Berke, A. Review: the future of artificial intelligence in strategic studies quarterly, Vol. 10, No. 3, Emerging technology special edition (Fall 2016), pp. 114–118, Air University Press, available at https://www.jstor.org/stable/10.2307/26271497, accessed on May 17, 2019.

[9] Burton, E., Goldsmith, J., Koenig, S., Kuipers, B., Mattei, N., Walsh, T. (2017). Ethical considerations in artificial intelligence courses. *AI Mag.*, **38**(2):22–34, doi: 10.1609/aimag.v38i2.2731, available at file:///C:/Users/user/Downloads/2731-Article%20Text-5314-1-10-20170603.pdf, accessed on May 17, 2019.

[10] Ibid. p-1.

governance" is needed in the matter of AI. AI has both positive and negative consequences for humanity, and whether human control on AI is indispensable needs to be considered. Therefore, it is a debatable issue whether matters related to AI require any supervision or control under any regulatory mechanism.

9.3 Ethics and Artificial Intelligence

In the modern world, the ongoing research on AI and robotics and its development have become a widely discussed topic. The issues of AI are not confined to the laboratories. They have started to affect our daily work qualitatively and quantitatively, both. The human-level intelligent artificial machines already have a grip on our routine works, and gradually, we are going to be more and more dependent on these as they perform faster and better. The Select Committee on Artificial Intelligence, in its Report of Session 2017–19, states that AI may prove beneficial to the British economy if it is nurtured properly. Not only that, it may also help the UK to solve various problems and improve its productivity.[11] This observation of the House of Lords clarifies the importance of AI in modern society. Undoubtedly, the knowledge revolution of the present era has generated huge data, and there has been an AI boom in order to have access these data. Researchers on AI claim that only AI can give specific and testable models for theories on memory, learning, language, and human inference.[12] In modern times, AI is going to affect our life even more than electricity and the Internet. AI performs the cognitive tasks performed by human beings. Modern machines not only mimic human intelligence but are bestowed with the power of creativity too. Progress from machine learning to deep learning is gradually enriching the systems of AI. A machine learning system works by analyzing the data given to it. This system works through a fixed algorithm and requires the assistance of a human, who will make necessary adjustments if the machine fails to provide the desired result. On the other hand, a deep learning

---

[11] Supra note 7, p-5.
[12] Supra note 5, p-6.

system classifies data on its own with a logic structure similar to the cognitive ability of the human mind. It works through the application of a layered structure of algorithms, technically named the artificial neural network. It studies the situation on its own, following the process of trial and error. An automatic car driving system is the best example of a deep learning process. It does not require assistance from a human agency afterward and can read the situation and may reach a conclusion similar to the cognitive power of the human mind. We can witness a double exponential growth of AI, that is, growth in the software system as well as in the hardware.

According to Allen Dafoe, who is the director of the AI program at the University of Oxford's Future of Humanity Institute, AI is the study of machines capable of sophisticated information processing.[13] He said in an interview with the Journal of International Affairs that by virtue of AI, we may automate, improve upon, or scale up critical human skills in prediction and decision making and this potential of AI makes it capable of bringing transformation into the fields of economy, society, and the military.[14] Andrew Ng, Google Brain Founder, had compared the present status of AI with that of the early state of electricity and said that just as the work of Thomas Edison led to the modern electric power–driven society, the time has come to make AI-powered society.[15] AI can be categorized into three kinds: artificial narrow intelligence (ANI), artificial general intelligence (AGI), and artificial super intelligence (ASI). ANI is called narrow intelligence because it performs a single task in real time and within a predefined or predetermined field, such as giving the weather forecast or playing chess. ANI does not have the consciousness, emotions, etc., of the humans.[16] On the other hand, AGI is considered to have strong intelligence, capable of acting as per the response of the cognitive faculty, as the human mind works,

---

[13] Dafoe, A. (Fall/Winter 2018). Global politics and the governance of artificial intelligence. *J. Int. Affairs*, **72**(1):121–126, available at https://www.jstor.org/stable/10.2307/26588347, accessed on Oct 20, 2019.

[14] Ibid. p-121.

[15] https://phys.org/news/2018-02-google-brain-founder-andrew-ng.html, accessed on Oct 26, 2019.

[16] https://medium.com/@tjajal/distinguishing-between-narrow-ai-general-ai-and-super-ai-a4bc44172e22, accessed on Oct 26, 2019.

and is also expected to be innovative, creative, and imaginative. AGI can solve problems logically, pass judgments with reason, and make decisions under uncertainty.[17] ASI is said to have surpassed human intelligence by many times in matters from creativity, to general wisdom, to problem solving.[18] Tesla CEO Elon Musk, in his debate with Alibaba cofounder Jack Ma on AI and its implications for humanity in the World Conference in Shanghai, opined that AI is not a smart human but it will be much smarter than the smartest human, which is beyond the imagination of even AI researchers.[19] People often confuse between robots and AI. Basically, a robot may be termed as the body of the structure while AI is considered as the mind of the structure. A robot may or may not be bestowed with the power of AI.

9.4 AI and Its Challenges

It is true that AI will bring radical transformation and structural changes in social, economic, political, security, judicial, education, transport, and health sectors. In other words, we can say that it will have an unimaginative but tremendous impact upon all facets of human life. AI will be expected to boost the gross domestic product by 1.2% per year by 2030.[20] AI researchers also claim that AI can successfully tackle global challenges like hunger, poverty, environmental pollution, earthquakes, deadly diseases, and real-time monitoring of crops risk to farmers and assistance to them to increase production.[21] According to Ma, AI will create the opportunity of lots of jobs, though he believes that in the AI period, people will not require many jobs. As per his prediction or assumption, in the AI period, people will be able to live for 120 years due to advances in medical facilities and the jobs that

---

[17] Ibid.

[18] Ibid.

[19] https://www.wired.com/story/elon-musk-humanity-biological-boot-loader-ai/, accessed on Oct 26, 2019.

[20] Posner, T., AI & Global Governance: artificial intelligence for all- a call for equity in the 4th industrial revolution, available at https://cpr.unu.edu/category/articles, accessed on Oct 15, 2019.

[21] Ibid.

people will require at that time will be mainly to do with making people happier, experiencing life, and enjoy it.[22] AI researchers are also of the opinion that it will bring tremendous opportunity for sustainable development because improved intelligence will help to minimize pollutants, check the misuse of resources, provide faster and more advanced information, supply environment friendly technologies, and ensure improved risk assessment systems and proper planning.

These positive aspects of AI notwithstanding, we cannot overlook its negative impacts, as often predicted by social scientists, researchers, and philosophers. AI has its own triumphs and turbulences, fears, expectations, and worries and a myriad of other challenges. The concern of ethical issues embedded within AI is growing with advancements in AI. Many questions and issues are springing up related to the AI regime, such as the impact of AI on human society, transparency, accountability and governance, changing legal and economic issues, security concerns, curtailment of jobs, the changing education scenario, and above all the role of human beings in the AI period. When the exponentially accelerating technology reaches the stage of AGI and ASI, it will exceed human intelligence by many times and will go beyond human control. These machines will make their own decisions and take independent steps. For its future course of action, AI will not depend on or wait for any instructions from humans. To Musk, the rate of advancement of computers is insane.[23] He said that the rate of change of technology is incredibly fast, outpacing our ability to understand it.[24] A technology guy like Musk is also confused about the consequences of such a situation.[25] Today, the triumph of machines over humans in a battle is not confined to fiction any more. The defeat of Gary Kasparov by Deep Blue in 2007 or of Lee Sedol by Alpha Zero reminds us the Puranic story of Bhasmasura, in which Lord Shiva, known as "The Destroyer," became pleased with the prayers of Bhasmasura and endowed him with superior powers and at the end of the episode was threatened with destruction by his own

---

[22] Supra note 20.
[23] Ibid.
[24] Ibid.
[25] Ibid.

creation. Here, ethics ask us to rethink our own capabilities, potentiality, and wisdom—are we really stepping prudently toward a safe future.

AI has posed some serious ethical challenges that need to be addressed by the world polity to save humanity and mankind. To Allan Dafoe, even if there was no further technical development of AI from today onward, the existing position of AI is sufficient to pose major governmental challenges.[26] Not a superhuman approach of AI, even a near-human performance of AI could lead to labor displacement; radically increased inequality; infringement of privacy; risk of nuclear instability; concentration of market power; structural changes in the military system; and enhancement of conflict on different issues, including ongoing research in cyber space.[27] It is said that the AI regime will badly affect the population. People will live longer and according to Ma, even great-grandfathers will be fit enough to work with their great-grandsons. Musk compared AI to the "call of the devil" and according to him, these uncontrolled machines may turn into "immoral dictators."[28] According to Musk, this superhuman or near-human approach of AI can be scarier than nuclear weapons because machines cannot be taught to be guided by either emotions or morality.[29] For example, there is talk about "killer robots" or "military robots" being developed and deployed by the United States, which have wheels and wings like the predator drones, are largely autonomous, and are remotely controlled by people from other parts of the world,[30] and there may be a surge in their production in the global arms race.[31] This innovation and its advanced technology in the near future may prove destructive for the whole humanity if somehow the driving force behind it goes into the possession of repressive governments, terrorist groups, or

---

[26] Supra note 14, p-121.

[27] Ibid. p-121–122.

[28] Abramovych, G. V., Ryndin, S. A. Implementation of artificial intelligence in human life, available at file:///C:/Users/user/Downloads/7008-24821-1-PB%20(4).pdf, accessed on Oct 15, 2019.

[29] Ibid.

[30] Keiper, A., Schulman, A. N. (Summer 2011). The problem with 'friendly' artificial intelligence. *New Atlantis*, **32**:80–89, available at https://www.jstor.org/stable/43152658, accessed on Oct 20, 2019.

[31] Supra note 10, p-4.

separatists.[32] AI has its own positive and negative consequences, and it is also true that we cannot stop the development of science, but what prevents us from applying our prudence and wisdom, which is our basic instinct and a unique gift of God embedded within us?

9.5 Issues of Human Rights, Governance, and AI

AI will be the order of the future, and we have already stepped into the AI regime. Still, before opening Pandora's box, we must take a last call, using our own consciousness and intelligence, regarding what we are going to install in AGI and ASI in the near future and also from our own conscience, which is our own unique property and which we cannot confer on any machine, what kind of society we are going to make for tomorrow. Before showing our smartness in creating superhuman AI, we must ensure that the whole process be governed from a humanistic approach based on values and human rights. It is indeed astonishing to know that only 0.004% of the global population has the knowledge and potential to build an AI system and control the important affairs of human life, such as the job market, the insurance sector, or the judicial organ.[33] The rule of law is the basic postulate of a democratic society, the edifice of which stands on transparency and equality. Arbitrariness is antithetic to equality, and it is always antagonistic to the humanistic system.

Since in the process of development of AI, distinct state actors, private actors, and multinationals are involved, it is important to know whether due diligence has been followed in the creation of software; what standard has been followed in such process; what code is written and by whom, when, and why; which software and data libraries are used; and what hardware is used during system development.[34] These conditions should be given legal sanctity and be considered as prerequisites for developing technology related

---

[32] Ibid.

[33] Pauwels, E. The ethical anatomy of artificial intelligence, available at https://cpr.unu.edu/category/articles, accessed on Oct 15, 2019.

[34] Bryson, J. No one should trust artificial intelligence, available at https://cpr.unu.edu/category/articles, accessed on Oct 15, 2019.

to AI. As designing of AI is confined in the hands of a few people across the world and the lion's share of the population either does not have any clear idea about AI or is not acquainted with this terminology at all, the few cannot be given full authority to make decisions and control the fate of the lion's share of the population. Here, it is pertinent to mention the Draft Recommendation of the Committee of Ministers to Member States on the Human Rights Impacts of Algorithmic Systems, 12 November, 2018, which clearly states in its preamble that as the process of algorithmic systems involves different actors, like designers, programmers, creators of algorithms, owners, sellers, users, and service providers (including private and public structure), the mechanism of cooperative responsibility and accountability, including the risk management process, is required to be established at all stages of the process in order to realize and promote public interest goals at all levels and to check the possible harms.[35] Not only that, uncontrolled development of technology may put unfettered power into the hands of a few to run the universe as per their whims, to crush the whole humanity, to ruin the democratic fabrics, and finally to break down the whole human society. The "right to know" has been embodied as a basic right to enjoy the "right to life," along with the fundamental freedom in the legal system of almost all democratic countries, including India. It is the need of the hour to create a legal structure to develop an environment of digital trust and cooperation so that the ongoing development of the AI mechanism is open to social scientists, technocrats, engineers, academicians, and activists of civil society and not a few of us but all of us collectively decide what kind of world we want to shape for tomorrow, where justice can be informed to all in the social, economic, and political fields. The prevailing situation related to AI does not prevent algorithms from infringing on fundamental human rights; rather the large store of data in the absence of a proper monitoring system may pose a serious threat to the "right to privacy," which is an important component of the "right to life." In

---

[35] Para 10 of the Preamble, available at https://rm.coe.int/draft-recommendation-on-human-rights-impacts-of-algorithmic-systems/16808ef256, accessed on Nov 08, 2019.

the absence of global governance, the installed code or algorithms may be guided by the devil mind and as a consequence, they may make a gender-biased, antiplural, nonsecular society. To overcome this untoward situation, the Preamble to the Draft Recommendation of the Committee of Ministers of Councils of European Union, 2018, further states that all possible measures are required to be taken to ensure that private actors engaged with the process of deep learning of the machine fulfill their responsibilities to respect human rights in tune with the guidelines of the Guiding Principles on Business and Human Rights Recommendation of United Nations and the guidelines of the European Council regarding the roles and responsibilities of Internet intermediaries.[36]

AI governance has to address both the foreseen as well as unforeseen challenges before it, such as programming related to values, including its manifold dimensions, to ensure accountability in automated action and decision, to provide democratic pace to the global citizens to have their say in development, etc. Whether it is superhuman or near-human intelligence, the global governance of AI must ensure that the driving force be possessed by the human being. To many researchers, the above is impossible to achieve because AI at the ASI stage will be self-conscious and much more powerful than the human being, but at the end of the hour, we should not forget that AI is the creation of human beings and now onward, our strength and wisdom should be guided with full vigor to create an alternative path to save humanity. Only cooperation among state actors, multinationals, individuals, and private actors can pave the way to maintain the integrity of the human society with a humanist approach. Here, it is pertinent to mention a number of reports from different world bodies focused on the issues of human rights in the context of the development of AI, for instance, the UN's "Special Rapporteur on the Promotion and Protection of the Right to Freedom of Opinion and Expression," the Council of Europe's "Draft Recommendation of the Committee of Ministers to Member States on the Human Rights Impacts of Algorithmic Systems," and Business for Social Responsibility's "Artificial Intelligence: A Rights Based

---

[36] Ibid. para 11 of the Preamble.

Blueprint for Business" series.[37] The Special Rapporteur on the Promotion and Protection of the Right to Freedom of Opinion and Expression, 2018, acknowledges that AI technologies have created tremendous opportunity for the freedom of speech and expression by sharing the information and ideas globally and quickly but at the same time it may tremendously interfere with the autonomy and self-determination of the people and it is a challenge before the world polity to ensure that AI technologies promote and respect human rights.[38] The above report recommends that AI technologies be designed in such a way that they are consistent with the obligations of states and responsibilities of private actors under the international human rights law.[39] It also reminds the states about their obligation under Article 19 of the International Covenant on Civil and Political Rights to compel the private actors to respect the opinions of individuals, who may take legal measures as and when required to attain the purposes in this respect.[40] The above report, in Paragraph 33, gives importance to the right to privacy as a large amount of data remains stored in the present AI system.[41] The report further points toward a glaring challenge in the AI system— the lack of provision for any effective remedy guaranteed by the human right laws to individuals determined by competent judicial, administrative, or legislative authorities;[42] and hence it suggests that the states adopt the strategies and relevant policies to uphold human rights.

---

[37] Pielemeier, J. The advantages of applying the international human rights frame-work to artificial intelligence, available at https://cpr.unu.edu/category/articles, accessed on Oct 15, 2019.

[38] Available at https://undocs.org/A/73/348, accessed on Oct 28, 2019.

[39] Ibid. para 19.

[40] Ibid. para 20.

[41] Ibid. para 33 states that the right to privacy often acts as a gateway to the enjoyment of the freedom of opinion and expression. 30 Article 17 of the covenant protects the individual against "arbitrary or unlawful interference with his or her privacy, family, home or correspondence" and "unlawful attacks on his or her honour and reputation" and provides that "everyone has the right to the protection of the law against such interference or attacks." The Office of the High Commissioner for Human Rights and the Human Rights Council have emphasized that any interference with privacy must meet standards of legality, necessity, and proportionality (A/HRC/27/37, para 23 and Human Rights Council resolution 34/7, para 2).

[42] Ibid. para 39 & 40.

The Draft Recommendation of the Committee of Ministers to Member States on the Human Rights Impacts of Algorithmic Systems directs the member states to promote and implement effective age- and gender-sensitive media to enjoy the opportunities and to minimize the harmful impact of algorithm systems in a communication network by engaging all the stakeholders involved within it,[43] and it has also imposed obligations on states that the legal mechanism related to the algorithmic system must safeguard human rights and fundamental freedom.[44] The above draft recommendation has also imposed obligation on the states to ensure that the legislations or regulations applicable to the algorithmic process be transparent, accountable, and inclusive and safeguard the interests of the public concerned.[45] It also asks for the creation of an expertise body to detect the cases related to human rights risks and to provide possible remedial actions[46] and also to evaluate and combat the bias and potential discriminating in the algorithmic processing.[47] The above recommendation also imposes duty on states to provide effective judicial and nonjudicial remedies to ensure impartial review of the process and to ensure justice.[48] The states should also conduct an impact assessment of the algorithmic process to check and mitigate the risks to human rights, incorporate dynamic testing methods, and compel the private actors to adopt the same process to check the probable incidents of violation of human rights.[49] The draft recommends that the state develop the algorithmic system in such a manner that it may create an enabling environment so that all can explore and enjoy human rights and fundamental freedoms.[50] The draft recommendation imposes an obligation on the private actors to collect data from reliable and diversified sources and regularly test the data against bias and for completeness, relevance, etc.[51] Through the above

---

[43] Supra note 37.
[44] Ibid. para 1.1 of 1of A.
[45] Ibid. para 1.6.
[46] Ibid. para 1.7.
[47] Ibid. para 2.4 of 2.
[48] Ibid. para 5.1.
[49] Ibid. para 6.3.
[50] Ibid. para 7.1.
[51] Ibid. para 2.1 of B part.

draft recommendation, the effort has been to try and install a safety valve in the process of the development of algorithms to protect human rights. Neither a public nor a private actor can be allowed to jeopardize the right to life and the fundamental freedom of people by developing algorithms capriciously, and therefore the above draft regulations rightly ask for the incorporation of a comprehensive legal system to provide justice to people against any unjust algorithm or data that undermine the integrity of humans. The most positive side of this draft regulation is that it advocates in favor of the creation of a body that consists of experts in the matter of AI to keep a vigil so that the development of technology cannot be used as a means of exploitation.

The apprehensions regarding human rights violations and the anxiety to protect human rights are gradually increasing with the development of the process of AI. Most of the human rights activists are not against the development of the system of AI, because the development of science and technology is normal and spontaneous and it is not possible to interrupt the above process for a long time, but they advocate developing the whole AI system from a humanistic approach so that the near-human approach or superhuman approach of AI system does not pose a threat to the existence of this—the wisest—creation of God. The Toronto Declaration, 2018, is a similar attempt of a group of human rights activists and technocrats to save humankind and give a specific orientation to uncontrolled technological development so that it may prove beneficial to human society. It called on both public and private actors to ensure that algorithms respect the right to equality and are based on the principle of nondiscrimination.[52] The main focus of the Toronto Declaration is on how to make technology humancentric and ensure that the issues related to AI, including its ethical portion, be judged through the human rights lens to assess present and potential future harm to human rights and also to take corrective measures in this regard to mitigate the risks,

---

[52] The Toronto declaration: protecting the right to equality and non-discrimination in machine learning systems, available at https://www.accessnow.org/cms/assets/uploads/2018/08/The-Toronto-Declaration_ENG_08-2018.pdf, accessed Nov 09, 2019. The above declaration was published by Amnesty international and accessed on May 16, 2018.

if any arises.[53] The above declaration advocates that under the international human rights laws, the states are bound to protect human rights, which include the rights not to be discriminated against, and keeping the above mantra in mind, the machine learning system should be designed in such a way that people can enjoy the right to life meaningfully, along with their right to privacy, fundamental freedom, etc. The machine learning system should be based on the principles of inclusion, diversity, and equity and must ensure transparency and accountability from the different actors associated with the algorithmic process.

9.6 Destiny of Humanity in the World of AI

Human rights activists are seriously thinking about the destiny of humanity in the world of AI. Even in this nascent stage of AI, the world is witnessing how an uncontrolled digital and algorithmic process can ravish the rights of the people. In Section 9.3, some of the issues have already been discussed, clarifying the depth of the problem. The growing use of AI in the justice delivery system may pose a threat to the integrity of the people, for example, the recidivism risk-scoring software widely used in the US criminal justice mechanism to make decisions on detainment issues, where the court decides everything, from the assignment of bail to criminal sentencing, with assistance from the above algorithm. But it has been found that the algorithm used in the above software is very harsh and detrimental to the black defendants. Along with the risk of falsely implicating the black respondents, the above algorithm imposes on them complicated bail conditions, pretrial detention, and harsh punishment.[54] Similarly, software like criminal risk assessment and predictive policing software, is found to be biased and discriminatory in nature.[55] Though these tools are made to assist judges, in reality, it has been found that many judges

---

[53] Ibid. para 8.

[54] Access now: human rights in the age of artificial intelligence available in https://www.accessnow.org/cms/assets/uploads/2018/11/AI-and-Human-Rights.pdf accessed on 06/11/2019.

[55] Ibid.

rely heavily on these software while making decisions without having any idea of how these software systems work. So, directly or indirectly, knowingly or unknowingly, private actors—basically some technocrats—are interfering with the justice delivery system and the delivery of judgment, which is alarming and suppresses human rights. The use of facial recognition software in the legal system raises the risk of unlawful arrest and detention as there is no dearth of identical persons or at least similarity among faces across the world.[56] Advancement of technology and the growth of the Internet system increases the surveillance system both in public and private spheres, and in modern times, it is easy even for an individual to closely monitor the movement and expressions of another from any other part of the globe. Almost all the information of an individual is available both in the public and the private domain, and these developments are seriously denting the right to privacy. The AI regime may seriously impair human rights, but the International Bill of Human Rights and Protocols contains a mechanism to save and protect the basic rights of the people. The "Guiding Principles on Business and Human Rights: Implementing the United Nations 'Protect, Respect and Remedy' Framework," developed by the Special Representative of the Secretary General and endorsed by the Human Rights Council, in its resolution 17/4 of June 16, 2011, can prove effective in protecting human rights in dealings of multinational corporations as well as other business enterprises, even in this digital era. It confers upon the states the duty to take appropriate steps to check and remedy the matters related to the violation of human rights through effective policies, legislation, regulations, and adjudication[57] and also ensures that business enterprises domiciled in their jurisdiction pay respect to human rights throughout their operations.[58] The above guiding principles direct both public and private actors to avoid violating human rights, and if such infringement does occur, the states are bound to take action as per the existing international laws on

---

[56] Ibid.
[57] Para 1 of Foundational principles, available at https://www.ohchr.org/ Documents/Publications/GuidingPrinciplesBusinessHR_EN.pdf, accessed on Nov 09, 2019.
[58] Ibid. para 2.

human rights. It clarifies that business entities, be it public or private, are bound to observe the existing laws and norms related to human rights and thus it obliges the corporate sector to follow the global ethics and standard related to human rights even on the threshold of AI. The business avocation should use its due diligence to detect the possible human rights risks and should not incorporate any policy statements that may have an adverse impact on human rights. The Guiding Principles on Business and Human Rights bestows on transnational corporations and other business enterprises the obligation to measure the potential adverse impact on human rights either by their own activities or as a consequence of their relationship and be involved in meaningful consultation with the potentially affected groups to chalk out the future operational courses.[59] It puts emphasis on the possible risk assessment in all sectors, including human rights impact assessment, by the avocation before or during business operations and these provisions may be equally applicable to concerns engaged in the formulations of algorithms or engaged in machine learning systems or later uses.

However, it is hard to read the composition of the society and the role of human beings with operations of human beings in the era of AGI and ASI. As of now, the violation of human rights, even in this digitalized system, can be effectively addressed under the existing human rights legal framework, provided there remains a strong political will; awareness of people; and humanistic policies based on justice, equity, and good conscience.

9.7 Laws and Policies Related to AI

Before preceding further, it is worthwhile to mention the "mantra" of the report of the Centre for European Policy Studies (CEPS), Brussels, in February 2019 titled "Artificial Intelligence Ethics, Governance and Policy Challenges," which states the AI is quite different from the earlier inventions, like the Industrial Revolution or the invention of the wheel or the introduction of the regime

---

[59] Ibid. para 18.

of the Internet, in that it is something special and the ultimate realization of the CEPS though AI is still in its infancy, and though it has a prominent presence in digital sectors, banking, e-commerce, insurance, healthcare, the energy sector, etc., it needs new laws, rules, and policies to interact with the new machines.[60] Nobody can rule out the fact that the AI regime will have a major impact on the global order and its balance of powers.[61] Vladimir Putin, in his speech, had pointed out that AI is a resource not only for Russia but for the whole mankind and whoever will lead in this sphere will become the leader of the world.[62] Perhaps to realize this above dream, almost all the big countries, like China, European Union, the United States, the UK, Russia, Japan, and India, have become involved in this rat race to become a global leader. China has already declared that it intends to become a world leader in AI mechanism by 2030.[63] India has also implemented dedicated strategies in this sphere in the year 2018.[64]

As far as the European Union is concerned, the High-Level Expert Group on Artificial Intelligence released a set of AI ethics to maximize the benefits and minimize the risk factors associated with it. Its main aim is to make AI humancentric and trustworthy. The EU is working on the reformation of its liability rules to enter into the AI regime. Currently, it is guided by two laws in this field, the Product Liability Directive and the Machinery Directive, which ensure health and safety both from the product and the machine, trying to balance between public interest and innovation in AI.[65] The UK emphasizes on sector-specific regulation in the matter of AI rather than the blanket AI-specific regulation. Here it is worthwhile to mention two regulations of the UK in this regard, the Data Protection Bill and the General Data Protection Rule, which may help to maintain the right to privacy.[66]

---

[60] Available at https://www.ceps.eu/wp-content/uploads/2019/02/AI_TFR.pdf, accessed on May 18, 2019.
[61] Ibid. p-37.
[62] Ibid.
[63] Ibid. p-38.
[64] Ibid.
[65] Ibid.
[66] Supra note 7, p-137.

In India, NITI Aayog has initiated a three-pronged approach to address the issues of AI: undertaking exploratory AI projects in different areas, creating a vibrant ecosystem, and collaborating with experts and stakeholders.[67] On June 4, 2018, it released a discussion paper "National Strategy on Artificial Intelligence," where it focuses on five sectors: healthcare, agriculture, education, smart cities, and transportation.[68] NITI Aayog decided to establish a comprehensive framework for data protection under the supervision of the Justice Srikrishna Committee and national data protection and privacy laws with international standards to protect the right to privacy,[69] which attained the status of a fundamental right in India, especially after the judgment of the Supreme Court of India in *Justice K.S. Puttaswamy (Retd.) and Anr. v. Union of India and Ors.*[70] The discussion paper of NITI Aayog has sought to impose liability and negligence test to make the involved parties accountable for their misdeeds. In India, a lack of expertise on AI, lack of an enabling data ecosystem, high resource costs, and the absence of regulations and collaborative approaches toward AI are the main impediments to achieving the goals of AI for all.[71] The above discussion paper proposes a two-tiered structure to achieve research in the field of AI, the creation of new knowledge through the Centre of Research Excellence and the creation of the International Centres of Transformational AI, the main motto of which is to develop and deploy application-based research in collaboration with the private sector.[72] In India, besides the Constitution—which is regarded as the supreme law of the land and ensures justice in social, economic, and political fields—and the already ratified Covenants on Civil and Political Rights and Covenants of Social, Economic and Cultural Rights, a number of acts have been enacted in different subjects and areas to protect

---

[67] Available at https://niti.gov.in/national-strategy-artificial intelligence#target Text=NITI%20Aayog%20after%20having%20round,of%20Sabka%20Saath% 20Sabka%20Vikas, accessed on Oct 28, 2019.

[68] Available at https://niti.gov.in/sites/default/files/2019-01/NationalStrategy-for-AI-Discussion-Paper.pdf, accessed on Oct 28, 2019.

[69] Ibid. p-87.

[70] Writ petition (civil) no. 494 of 2012, available at https://indiankanoon.org/ doc/127517806/, accessed on Oct 28, 2019.

[71] Supra note 70, p-7.

[72] Ibid.

human rights, such as the Protection of Human Rights Act, 1993, which has paved the way for the creation of national as well as state human rights commissions to provide remedy to the people against state tyranny and any kind of oppression. In India now, the right to privacy is a fundamental right. It is the cardinal duty of the state to impose checks on unregulated and arbitrary use of data so as to protect the privacy and autonomy of an individual. Here, it is worthwhile to mention Section 43 of the IT Act, 2008, which enjoins the body corporate that deals with sensitive personal data to provide compensation to the affected party if any wrongful gain or wrongful loss occurs to any person due to the negligence of such a body corporate in implementing and maintaining a reasonable security practice. Information Technology (Reasonable Security Practices and Procedures and Sensitive Personal Data or Information) Rules, 2011, states that sensitive personal data or information refers to such personal information that consists of the information related to passwords, financial information, and state of health (including medical records, sexual orientation, and biometric information) but does not include any such information that is already in the public domain or can be supplied under the Right to Information Act, 2005.[73] However, it is pertinent to mention that in the Justice Puttaswamy case, it has been clarified by the Supreme Court of India that no citizen will be compelled to provide his or her Aadhar card, which consists of, besides other things, the biometric information of the person. Under this rule, the body corporate before collecting any sensitive personal data or information must allow the provider of information the option not to supply the information as asked for and to withdraw his or her consent at any point of time[74] and disclosure to a third party of any such information requires the consent of the provider of information.[75] Such a body corporate has to designate a grievance officer to redress the grievances of the people in a time-bound manner[76] and must adopt the reasonable security practices and procedures of the International standard

---

[73] Sec. 3 of the rules, 2011, available at https://indiacode.nic.in/, accessed on Nov 12, 2019.
[74] Ibid. rule 5(7).
[75] Ibid. rule 6.
[76] Ibid. rule 5(9).

IS/ISO/IEC 27001.[77] However, we don't have any stringent and comprehensive laws regarding data protection and though the Personal Data Protection Bill, 2018, was introduced in the year 2018, it has not yet attained the status of an act.

The state actors across the globe are trying to formulate policies and legal structures to welcome the AI regime, but AI is still in its nascent stage and it is impossible to foresee all the future consequences. It is a sad state of affairs that the states— mainly the developed countries—are competing with each other to take leadership in an AI-controlled globe instead of extending cooperation to each other to deal with the uncertain situation. The world polity is not coming ahead to draw any blueprint for imposing reasonable restrictions on the algorithms to save mankind with a humanist approach. The UN should show a more pragmatic approach and positive actions in the matter of AI. However, the advisory bodies of UNESCO have already prepared various reports, like the Universal Declaration on the Human Genome and Human Rights, 1997; the Universal Declaration on the Human Genome and Human Rights, 1997; and the Report of the World Commission on the Ethics of Scientific Knowledge and Technology on Robotics Ethics, 2017, which can help the humanity to overcome the adverse situation created by a supercomputer age.[78]

9.8 Concluding Remarks

It will be fitting to conclude with Roger Penrose's prologue in his famous book *The Emperor's New Mind: Concerning Computers, Minds and the Laws of Physics*, where in the inauguration program of a supercomputer that could resolve any complex problem within the fraction of a nanosecond, the audience was asked to put the first question before the computer. Scientists, mathematicians, engineers, political personalities, bureaucrats, and of course the creator of this supercomputer were present within the audience.

---

[77] Ibid. rule 8.
[78] Azoulay, A. Artificial intelligence: humanity's new frontier, available at https://www.un.org/en/chronicle/article/towards-ethics-artificial-intelligence, accessed on June 05, 2019.

Everyone among the audience started to look at one another in dismay because no one wanted to look foolish and to expose his or her ignorance by putting a question before a machine endowed with superpowers. Then suddenly, a child came forward to put a question. Everybody smiled at the innocence of the child, but then the child asked, "How does it feel to be a computer?" The lights of the computer started to blink to respond but finally after more than one hour, the answer that flashed on the screen was, "I don't know."[79] This story signifies that nothing can be more powerful than human wisdom and consciousness. Ma, in his debate with Musk, referred to earlier in this chapter, made a significant remark that we should not worry about machines. Man cannot create even a mosquito, let alone a man. A computer is only that—a computer with a chip. On the other hand, a man has a heart, from where comes wisdom. So ultimately, human beings will win.[80] Though Ma is in favor of stepping into the world of AI, perhaps it is the need of the hour to take assistant of our conscience in this regard.

Further Readings

1. D'Alarcao, H. (1984). Artificial intelligence: myths and realities. *Bridgewater Rev.*, **3**(1):13–16, available at http://vc.bridgew.edu/br_rev/vol3/iss1/7.

2. Goldberg, S. (1994). Artificial intelligence and the essence of humanity, p-152 in the book entitled *Culture Clash*, NYU Press, available at https://www.jstor.org/stable/j.ctt9qfqbm.12.

3. Dreyfus, H. L. Artificial intelligence in the annals of the American Academy of Political and Social Science, Information Revolution, March 1974, p-22 quoted from Simon, H. A., Newell, A. (1958). Heuristic problem solving: the next advance in operation research. *Oper. Res.*, **6**(1):1–10, available at https://www.jstor.org/stable/1040396.

4. David Bolter, J. (1984). Artificial intelligence. *Daedalus*, Vol. 113(3), Anticipations (Summer, 1984), p-1 quoted from Semantic Information

---

[79] Atmapriyananda, S. in his speech on the science of Life and Living Swami Vivekananda's Worldview in ICAR quoted the story of Roger Penrose, available at https://icar.org.in/files/Vichar-29-Print.pdf, accessed on Oct 24, 2019.
[80] Supra note 52.

Processing, by Minsky, M., MIT Press, Cambridge, 1968), available at https://www.jstor.org/stable/20024925.

5. Report of the House of Lord, Select committee on artificial intelligence of session 2017-19, p-14 quoted originally from Department for Business, Energy and Industrial Strategy, *Industrial Strategy: Building a Britain fit for the future* (November 2017), p. 37, available at https://publications.parliament.uk/pa/ld201719/ldselect/ldai/100/100.pdf.

6. Berke, A. (2016). Review: the future of artificial intelligence in strategic studies quarterly, Vol. 10(3), Emerging technology special edition (Fall 2016), pp. 114–118, Air University Press, available at https://www.jstor.org/stable/10.2307/26271497.

7. Burton, E., Goldsmith, J., Koenig, S., Kuipers, B., Mattei, N., Walsh, T. (2017). Ethical considerations in artificial intelligence courses. *AI Mag.*, **38**(2):22–34, 10.1609/ aimag.v38i2.2731, available at file:/// C:/Users/user/Downloads/2731-Article%20Text-5314-1-10-20170 603.pdf.

8. Dafoe, A. (Fall/Winter 2018). Global politics and the governance of artificial intelligence. *J. Int. Affairs*, **72**(1):121–126, The fourth industrial revolution, available at https://www.jstor.org/stable/10.2307/ 26588347.

9. Posner, T. AI & global governance: artificial intelligence for all- a call for equity in the 4th industrial revolution, available at https:// cpr.unu.edu/category/articles.

10. Abramovych, G. V., Ryndin, S. A. Implementation of artificial intelligence in human life, available at file:///C:/Users/user/Downloads/7008-24821-1-PB%20(4).pdf.

11. Keiper, A., Schulman, A. N. (Summer 2011). The problem with 'friendly' artificial intelligence. *New Atlantis*, **32**:80–89, available at https://www.jstor.org/stable/43152658.

12. Pauwels, E. The ethical anatomy of artificial intelligence, available at https://cpr.unu.edu/category/articles.

13. Bryson, J. No one should trust artificial intelligence, available at https://cpr.unu.edu/category/articles.

14. Para 10 of the Preamble, available at https://rm.coe.int/draft-recommendation-on-human-rights-impacts-of-algorithmic-systems/ 16808ef256.

15. Pielemeier, J. The advantages of applying the international human rights framework to artificial intelligence, available at https://cpr. unu.edu/category/articles.

16. The Toronto declaration: protecting the right to equality and non-discrimination in machine learning systems, available at https://www.accessnow.org/cms/assets/uploads/2018/08/The-Toronto-Declaration_ ENG_08-2018.pdf, accessed 09-11-2019, Amnesty International and Access Now.

17. Azoulay, A. Artificial intelligence: humanity's new frontier, available at https://www.un.org/en/chronicle/article/towards-ethics-artificial-intelligence.

18. Atmapriyananda, S. in his speech on the science of Life and Living Swami Vivekananda's Worldview in ICAR quoted the story of Roger Penrose, available at https://icar.org.in/files/Vichar-29-Print.pdf.

19. Access Now, Human rights in the age of artificial intelligence, available at https://www.accessnow.org/cms/assets/uploads/2018/11/AI-and-Human-Rights.pdf accessed.

20. https://niti.gov.in/sites/default/files/2019-01/NationalStrategy-for-AI-Discussion-Paper.pdf.

21. Writ petition (Civil) No. 494 of 2012, available at https://indiankanoon.org/doc/127517806/.

22. https://phys.org/news/2018-02-google-brain-founder-andrew-ng.html.

23. https://medium.com/@tjajal/distinguishing-between-narrow-ai-general-ai-and-super-ai-a4bc44172e22.

24. https://www.wired.com/story/elon-musk-humanity-biological-boot-loader-ai/.

25. https://undocs.org/A/73/348.

26. https://www.ceps.eu/wp-content/uploads/2019/02/AI_TFR.pdf.

27. https://www.ohchr.org/Documents/Publications/GuidingPrinciples BusinessHR_EN.pdf.

28. https://niti.gov.in/national-strategy-artificial intelligence#targetText= NITI%20Aayog%20after%20having%20round,of%20Sabka%20Saath %20Sabka%20Vikas.

29. https://indiacode.nic.in/.

Index

Printed in the United States
by Baker & Taylor Publisher Services

Printed in the United States
by Baker & Taylor Publisher Services